String Theory
Weaving the Fabric of the Cosmos

Copyright © 2024 by Nobtrex LLC
All rights reserved. No part of this publication may be reproduced, distributed, or transmitted in any form or by any means, including photocopying, recording, or other electronic or mechanical methods, without the prior written permission of the publisher, except in the case of brief quotations embodied in critical reviews and certain other noncommercial uses permitted by copyright law.

Contents

1 **Introduction to String Theory: The Basics**   **13**
    1.1   What is String Theory?   13
    1.2   The Birth of String Theory: A Brief History   16
    1.3   Core Principles of String Theory   18
    1.4   Simple Analogies to Understand String Theory   20
    1.5   Comparing String Theory with Classical Physics   22
    1.6   Fundamental Components: Exploring Strings   25
    1.7   How String Theory Resolves Particle Physics Dilemmas   27
    1.8   Understanding the Significance of Vibrating Strings   30
    1.9   String Theory's Answer to Quantum Gravity   32
    1.10   String Theory Vs. General Relativity: A Comparison   35

2 **Historical Evolution of String Theory**   **39**

- 2.1 Early Theoretical Ideas Paving the Way for String Theory 40
- 2.2 The First Formulation of String Theory in the 1960s 42
- 2.3 The Renaissance of String Theory in the 1980s 44
- 2.4 Development of the Five Consistent String Theories 47
- 2.5 Introduction of D-branes in the 1990s 49
- 2.6 The Second Superstring Revolution and Dualities 52
- 2.7 The Era of M-theory and the Theory of Everything 54
- 2.8 Impact of String Theory on High-energy Physics 56
- 2.9 String Theory in the 21st Century: Enhancements and Impact 59
- 2.10 Major Figures in the Development of String Theory 61

3 Fundamental Concepts: Strings, Branes, and Extra Dimensions   65

- 3.1 Defining Strings: The Basic Building Blocks   66
- 3.2 Types of Strings: Open and Closed Strings .   68
- 3.3 Introduction to Branes: Multi-dimensional Spaces 70
- 3.4 Different Types of Branes: From D-branes to M-branes 72
- 3.5 Concept of Extra Dimensions: Beyond the Familiar Three 75

- 3.6 Compactification: How Extra Dimensions are Hidden 77
- 3.7 The Role of Strings and Branes in Higher Dimensions 79
- 3.8 Vibrational Patterns of Strings and Their Implications 82
- 3.9 Interactions Between Strings and Branes .. 84
- 3.10 Theoretical Predictions from Extra Dimensions 87

4 Mathematical Underpinnings of String Theory 91
- 4.1 Overview of Essential Mathematical Tools . 92
- 4.2 Calabi-Yau Spaces and Their Role in String Theory 94
- 4.3 Topology and Manifolds: Setting the Stage for Extra Dimensions 96
- 4.4 Conformal Field Theory in String Dynamics 99
- 4.5 Supergravity in Higher Dimensions 102
- 4.6 Algebraic Geometry: Essential Techniques for Brane Construction 104
- 4.7 Quantum Geometry and Its Implications for String Theory 107
- 4.8 Supersymmetry: Bridging Mathematics with Physical Phenomena 109
- 4.9 Dualities and Symmetries: Mathematical Magical Mirrors 112
- 4.10 Applying String Theory Mathematics to Other Fields of Physics 115

5 Quantum Mechanics and String Theory: A Symbiotic Relationship — 119

- 5.1 Basics of Quantum Mechanics: A Quick Overview 120
- 5.2 Integrating Quantum Mechanics with String Theory 122
- 5.3 Quantum Field Theory and Its Connection to Strings 125
- 5.4 The Role of Quantum Mechanics in Describing String Interactions 127
- 5.5 Quantum Entanglement and Superposition in String Theory 130
- 5.6 Quantum Fluctuations and String Dynamics 132
- 5.7 Non-locality and Locality Perspectives in String Theory 134
- 5.8 Quantum Gravity: The Joint Venture of String Theory and Quantum Mechanics .. 137
- 5.9 Holographic Principle: A Revolutionary Insight 140
- 5.10 Challenges and Opportunities in Merging Quantum Mechanics with String Theory .. 142

6 Key Experiments and Observational Evidences — 147

- 6.1 The Challenge of Testing String Theory .. 148
- 6.2 Historical Experiments Influencing String Theory Development 150
- 6.3 Observational Evidence for Extra Dimensions 152

CONTENTS

- 6.4 Methods of Detecting Vibrational Patterns of Strings 154
- 6.5 Utilizing Particle Accelerators: Exploring String Theory Predictions 157
- 6.6 Gravitational Wave Detection and Implications for String Theory 160
- 6.7 Cosmological Observations: Insights into Branes and Extra Dimensions 162
- 6.8 Microscopic Black Holes and String Theory Tests 164
- 6.9 Role of String Theory in Explaining Quantum Phenomena 167
- 6.10 Future Experiments and Technologies to Test String Theory 170

7 String Theory and the Nature of Reality: Space, Time, and Matter — 173

- 7.1 Redefining Space and Time through String Theory 174
- 7.2 The Concept of Matter in String Theory .. 176
- 7.3 How String Theory Modifies Our Understanding of Gravity 179
- 7.4 The Relationship between Vibrating Strings and Fundamental Particles 181
- 7.5 Space-Time Fabric and Membranes in String Theory 184
- 7.6 Emergent Phenomena: Space-Time from a Network of Strings 186

- 7.7 Implications of Extra Dimensions on Reality Perception 188
- 7.8 Time Dilation and Length Contraction: Relativity Meets String Theory 191
- 7.9 Quantum Coherence and Decoherence in String Theory 193
- 7.10 Exploring the Multiverse: Realities Beyond our Own 195

8 Unification and The Grand Design: Linking Forces and Particles — 199

- 8.1 The Quest for Unification in Physics 200
- 8.2 String Theory's Approach to Unifying Forces 203
- 8.3 Electromagnetism and Weak Force: Unification through Strings 205
- 8.4 Strong Force Integration in the Context of String Theory 208
- 8.5 Incorporating Gravity: The Ultimate Challenge 210
- 8.6 Supersymmetry: Balancing the Particle Universe 213
- 8.7 From Particles to Strings: Transitioning the Fundamentals 215
- 8.8 Predictions of String Theory on Particle Interactions 218
- 8.9 Experimental Signs of Unification from Particle Accelerators 220
- 8.10 The Role of Extra Dimensions in Force Unification 223

CONTENTS

9 Challenges and Critiques of String Theory   **227**

   9.1   Overview of Major Criticisms of String Theory 228

   9.2   The Problem of Testability and Falsifiability 229

   9.3   Mathematical Complexity and Conceptual Understandability 232

   9.4   The Landscape Problem: Multitude of Possible Solutions 235

   9.5   Dependence on Higher Dimensions: A Double-Edged Sword 237

   9.6   Lack of Unique Predictions: Challenges in Empirical Verification 240

   9.7   Competing Theories: Loop Quantum Gravity and Others 242

   9.8   Philosophical Implications: Science or Philosophy? 245

   9.9   Evolving Theories: Adaptability and Modification of String Theory 248

   9.10 Future Paths: Addressing the Critiques Effectively 250

10 The Future of String Theory: Theoretical and Experimental Prospects   **255**

   10.1 Current State and Immediate Future Directions in String Theory 256

   10.2 Advancements in Mathematical Formulations and Techniques 258

   10.3 Emerging Technologies and Their Impact on Experimental String Theory 261

10.4 Integration with Quantum Computing and Information 263

10.5 Colliding Worlds: String Theory Meets Phenomenology 266

10.6 Prospects for Uncovering Extra Dimensions 268

10.7 New Generations of Particle Accelerators and Detectors 271

10.8 Interdisciplinary Approaches: From Cosmology to Condensed Matter 274

10.9 Growing Global Collaborations and Projects in String Theory 276

10.10 Long-term Outlook: A Vision for String Theory in the Next Century 279

Preface

String Theory: Weaving the Fabric of the Cosmos is a comprehensive guide designed to delve deep into the intricate world of string theory, aiming to bridge the gap between complex theoretical physics and an intellectually curious audience with a collegiate level of understanding. This book is meticulously structured to unfold the fundamental elements and advanced concepts of string theory, exploring its developmental history, mathematical foundations, and its profound implications on our understanding of the universe.

The objective of this book is threefold. Firstly, it seeks to introduce the readers to the basic notions of string theory, providing a solid foundation of the subject matter without assuming prior advanced knowledge of theoretical physics. Secondly, it aims to illuminate the relationship between string theory and other areas of physics, articulating how this theory attempts to unify the different forces and particles within the universe under a single theoretical framework. Finally, the book strives to present a critical review of the theory, discussing its achievements and challenges, and offering insights into potential future developments and experimental prospects.

The substance of the book is organized into carefully curated chapters, each focusing on specific aspects of string theory. From the historical evolution of the theory and its fundamental concepts—including strings, branes, and extra dimensions—to its mathematical underpinnings and implications in quantum mechanics, the book provides an extensive coverage of the subject. Each chapter is designed to build upon the information presented previously, facilitating a gradual and comprehensive understanding of the topic.

Targeted at readers with at least a college-level education but not necessarily specializing in physics, this book aims to cater to anyone with a keen interest in modern theoretical physics and cosmology. Whether a student, an educator, or a passionate amateur, the reader will find this book an invaluable resource for gaining a clear understanding of string theory and its role in modern physics.

Through rigorous exploration and clear exposition, String Theory: Weaving the Fabric of the Cosmos offers a portal into one of the most fascinating areas of contemporary physics. It invites readers to explore the depths of theoretical physics, uncovering the beauty and complexities of the fabric of our cosmos.

Chapter 1

Introduction to String Theory: The Basics

This chapter lays the foundation for understanding string theory, beginning with a clarification of the core principles that define the theory. It traces the origins, highlights simple yet powerful analogies for easier comprehension, and compares string theory to classical physics constructs. Essential components such as strings, their attributes, and their significance in resolving fundamental particle physics dilemmas are discussed. The chapter concludes by exploring the overarching significance of these vibrating strings and how they provide insights into quantum gravity and the relationship with general relativity.

1.1 What is String Theory?

At its heart, String Theory is a framework in theoretical physics where the point-like particles of particle physics are replaced by one-dimensional objects known as strings. These strings can vibrate at different frequencies, and each vibration mode represents a different particle. Thus, String Theory aspires to be a unified theory of the universe, reconciling both general relativity (gravity) and quantum physics.

The core idea behind String Theory is that the myriad of observed particles are not, in fact, points, but rather small, oscillating strands of energy. The vibrational patterns of these strings determine their physical properties, such as mass, charge, and spin. Unlike in particle physics, where particles are idealized as points without any dimensions, strings have a finite length but negligible width, an attribute that aids in resolving various theoretical problems, such as the renormalization of gravity.

Moreover, these strings operate in a space-time that encompasses more dimensions than the 3+1 (three spatial dimensions and one temporal dimension) we are accustomed to in our daily experiences. Most versions of String Theory require a universe with 10 or 11 dimensions—the additional dimensions being 'compactified' or curled up into sizes too small to be observable at human scales. The manner in which these extra dimensions are configured gives rise to different solutions to string equations and, by extension, potentially different universes in a multiversal framework, highlighting the deep implications of the theory.

Mathematically, strings can be open (having two distinct endpoints) or closed (forming a continuous loop). The dynamics and interactions of these strings are described by the complex equations of String Theory, using techniques from quantum field theory, general relativity, and complex mathematics.

1.1. WHAT IS STRING THEORY?

An intriguing aspect of String Theory is its reliance on supersymmetry, a principle that posits each particle has a superpartner differing by half a unit of spin. This requirement stems from the necessity to include fermions (the building blocks of matter) and bosons (force carriers) within a unified framework in an asymmetrical yet harmonized manner.

String wave function: $\Psi(\text{String Position}) = \exp(i \cdot (k \cdot X - \omega t))$

Above, k denotes the wave vector representative of spatial frequency while ω signifies angular frequency correlating with time. This wave function shows the quantum nature of strings, portraying how their vibrations adhere to principles of quantum mechanics.

To better illustrate the concept, consider how a violin string can produce distinct musical notes based on vibrational patterns. Analogously, strings under String Theory oscillate to produce particles with specific attributes—a quark, a lepton, or a photon can all result from strings vibrating in different modes. Each mode, defined by string theory's equations, contributes uniquely to the universe's symphony of fundamental particles.

As we continue to discuss String Theory through the subsequent sections, we'll delve deeper into how these small strings shape the vast canvas of the cosmos through their vibrational harmonics. Combining the insights from this discussion with forthcoming topics will further illuminate the significance of strings in not only conceptualizing the microcosmic underpinnings of reality but also in addressing some of the most profound questions of our macrocosmic existence.

1.2 The Birth of String Theory: A Brief History

The inception of string theory can be traced back to the late 1960s, during an era dominated by a quest to understand the fundamental forces and constituents of matter. It was initially conceived not as a theory of everything but rather as a framework for describing the strong nuclear force. This precursor to string theory, known as the "dual resonance model," was proposed by physicists including Gabriele Veneziano, who sought a mathematical description for the forces that bind quarks together inside atomic nuclei.

Veneziano's model, introduced in 1968, provided an elegant solution for calculating the probabilities of various scattering processes involving strong interactions, using a formula originally formulated for the scattering of hadrons. Intriguingly, this formulation implied the existence of a one-dimensional object rather than point-like particles, inadvertently laying the groundwork for string theory.

As the theory evolved, it underwent significant transformations, particularly in the mid-1970s, a pivotal moment that reshaped its destiny. Holger Bech Nielsen and Leonard Susskind independently proposed that the fundamental components in Veneziano's model could be conceptualized as one-dimensional strings rather than zero-dimensional points. This shift from point particles to strings opened up new vistas for addressing unresolved issues in particle physics and beyond.

The real turning point came in 1974 when John Schwarz and Joel Scherk, and independently Tamiaki Yoneya, proposed that the strings in these models could be interpreted not only as mediators of the strong force but as fundamental constituents of all matter and forces, including gravity. Their pioneering work suggested that string theory could offer a unified description of all forces and particles, setting the stage for its recognition as a candidate for a Theory of Everything.

The late 1970s and early 1980s witnessed a burgeoning interest in string theory, underscored by the development of what is now known as the first superstring revolution. During this period, theorists discovered that string theory naturally incorporates supersymmetry, a mathematical symmetry that relates bosons and fermions. This revelation was critical because it allowed string theory to avoid certain theoretical inconsistencies and predicted new relationships between matter and forces.

The excitement was further amplified in 1984, when Michael Green and John Schwarz demonstrated that anomalies, mathematical inconsistencies that could potentially invalidate the theory, could be canceled out in string theory if certain conditions were met. This discovery led to an explosion of interest in string theory, marking the beginning of the second superstring revolution. Researchers delved into the five different versions of string theory that had been developed—Type I, Type IIA, Type IIB, SO(32) heterotic, and $E_8 \times E_8$ heterotic—each providing a unique yet interconnected way of describing the universe's fabric.

This period of intense theoretical exploration laid the foundational elements of what is now embraced globally as a robust framework attempting to reconcile quantum mechanics and general relativity, two pillars of modern physics often seen at odds at fundamental levels. The evolution from a niche model describing the strong nuclear force to a leading candidate for unifying all of nature's fundamental forces epitomizes the dynamic and revolutionary character of string theory.

The trajectory of string theory from its nascent stages to its current status illuminates a remarkable journey of intellectual discovery and innovation. The historical evolution underscores not only the scientific progress but also the persistent endeavor of the physics community to delve deeper into the underlying principles shaping our universe. By examining this progression, one gains a profound appreciation of both the achievements and ongoing challenges within theoretical physics, continually driving our understanding toward unprecedented horizons.

1.3 Core Principles of String Theory

In the quest to comprehend the universe at its most fundamental level, string theory stands out as a profoundly significant framework. Unlike point-like particles of particle physics, string theory posits that the basic constituents of reality are not zero-dimensional points, but rather one-dimensional strings. These strings can vibrate at different frequencies, and the mode of vibration determines the type of particle the string represents. Thus, at the core of string theory lies the principle that all matter and force particles are manifestations of these oscillating strings.

The strings in question can be open (having two distinct

endpoints) or closed (forming a continuous loop). The behavior of these strings, including their interactions, splitting, and recombining, is subject to quantum mechanics. The amplitude for a string to transition from one state to another – crucially, from one mode of vibration to another – is computed by path integral approaches similar to those used in quantum field theory.

A central tenet of string theory is holography, particularly the correspondence between theories formulated in different numbers of dimensions. The most celebrated example is the Anti-de Sitter/Conformal Field Theory (AdS/CFT) correspondence, which conjectures an equivalence between a string theory formulated in an AdS space and a quantum field theory formulated in one less spatial dimension.

Another core principle is supersymmetry, a proposed symmetry of nature that predicts that each particle has a corresponding supersymmetric partner, differing by half a unit of spin. Through this framework, string theory aims to unify all fundamental forces of nature, including gravity, within a single theoretical framework. This unification is pursued under various string theories such as Type I, Type IIA, Type IIB, and Heterotic string theories which, despite their differences, share the common features of incorporating supersymmetry and extra dimensions.

String theory also inherently includes the existence of extra spatial dimensions in addition to the commonly observed three spatial and one-time dimension. These extra dimensions are typically compactified on complex shapes known as Calabi-Yau manifolds, which are not directly observable but whose geometry affects the types of vibration possible for strings and hence the properties of fundamental particles.

Despite its profound implications, string theory is mathematically demanding and experimentally unverified in certain contexts. It requires the formulation of its principles and predictions in higher-dimensional spaces, and then meticulous work to draw phenomenologically relevant conclusions about our observable four-dimensional universe. Yet, its capability to offer insights into both particle physics and cosmology marks it as an indispensable area of modern theoretical physics.

Let us delve deeper into how these strings function dynamically and interactively in the next segments, enriching our understanding of the universe's fabric through the unique lens of string theory.

1.4 Simple Analogies to Understand String Theory

To demystify the complex concepts of string theory, it can be beneficial to draw parallels with familiar everyday experiences. Analogies, although not perfect, assist in transcending the boundaries of highly abstract ideas, making them more palatable for those not specialized in the field of theoretical physics.

One primary analogy used to explain the fundamentals of string theory is the comparison of subatomic particles to tiny, vibrating strings. Imagine a guitar string. When plucked, it vibrates and creates a note. Each note's pitch depends on several factors — the string's tension, its length, and its mass. Similarly, in string theory, the universe's fundamental particles are envisioned as incredi-

1.4. SIMPLE ANALOGIES TO UNDERSTAND STRING THEORY

bly small strings. These strings vibrate at different frequencies, and it's these frequencies that dictate the strings' different properties — such as mass and charge — just as different vibrations of a guitar string produce different notes.

This analogy extends further when considering harmonics. A guitar string can produce a fundamental note and several harmonics, or overtones, each corresponding to a distinct vibrational pattern. In a comparable fashion, a single fundamental string in string theory can manifest as different particles based on its mode of vibration. Thus, just as pressing down on a guitar string at various frets yields different tones, altering the vibrational state of strings in string theory results in different particles. This elegant model simplifies the vast array of particles observed in nature into variations on a singular 'string' theme.

Another useful analogy involves comparing the universe's makeup in string theory to a symphony orchestra. In an orchestra, each type of instrument has a unique sound, yet when played together under the conduction of a composed score, they create a harmonious melody resonating greater than the sum of its parts. String theory posits a similar idea where each vibrating string (like an instrument) contributes its unique 'note' or particle manifestation to the universe's grand composition, culminating in the complex world we perceive.

Further, one can think of the fabric of spacetime as akin to a fishnet, stretched taut. When objects with mass, such as planets or stars, press into this net, they create depressions or dimples. Much like how a heavy stone placed in the center of a stretched sheet will cause the sheet to warp, so too do these cosmic bodies bend the structure of spacetime, an effect we perceive as gravity. In string theory, this fabric is vibrational and dynamic, with strings influencing the fabric's properties at the most minute levels.

To visualize multidimensional spaces involved in string theory — as there are proposed extra dimensions beyond our observable three-dimensional space — one might think of higher-dimensional spaces as a series of interlocking rooms, each accessible through doors that one can only find and open using very specific keys. These keys in the universe are energies at scales must higher than what we currently observe. Effectively, while we live our lives largely unaware of them, these extra dimensions interact subtly with the dimensions we are aware of.

Each of these analogies helps in stepping away from the abstraction inherent to string theory, giving us models through which we can somewhat relate to or visualize profoundly un-intuitive concepts. Through familiar phenomena — vibrations, musical harmonies, or even simple fishing nets — we discover pathways leading towards grasping the otherwise elusive fabric of our cosmos as theorized in string theory. Although simplifications, these analogies serve as crucial tools to aid in teaching, understanding, and further contemplating this intriguing and potentially unifying framework in physics.

1.5 Comparing String Theory with Classical Physics

To effectively anchor our understanding of string theory, it is crucial to draw comparisons with classical physics—the familiar bedrock upon which much of mod-

1.5. COMPARING STRING THEORY WITH CLASSICAL PHYSICS

ern physics is built. Classical physics, epitomized by the works of Newton and Maxwell, describes the mechanics of visible objects and electromagnetic waves in terms of particles moving through space and time. In contrast, string theory posits that these apparently point-like particles are actually one-dimensional "strings" that vibrate at different frequencies, each mode corresponding to a different type of particle.

The fundamental divergence between classical physics and string theory begins with the concept of the atom itself. While classical physics views electrons, protons, and neutrons as point particles, string theory asserts that these are actually minuscule loops or strings. Each vibrating string produces distinct particles, depending on its vibrational mode, suggesting an elegant unification of force and matter under a single theoretical framework.

However, the uniqueness of string theory extends beyond its fundamental constituents. Classical physics is firmly rooted in four dimensions—three of space and one of time. String theory, on the other hand, proposes the existence of additional spatial dimensions, which are compactified or wrapped up in such a manner that they are imperceptible at ordinary scales. This postulation springs from the requirement that string theory's equations remain consistent and free from anomalies.

The methods used to derive predictions from the theories also starkly contrast. Classical physics often utilizes straightforward mathematical models that can directly correspond with experimental data. Conversely, string theory's mathematics is far more complex, requiring advanced techniques from fields like differential geometry, quantum field theory, and complex analysis to solve its foundational equations.

Further distinguishing itself from classical physics, string theory introduces a new perspective on gravity. Newton's universal law of gravitation and Einstein's general relativity describe gravity as a force acting across spacetime, influenced directly by the mass of objects. String theory interprets gravity as the manifestation of a particular vibrational mode of strings—specifically, those corresponding to hypothetical graviton particles. This seamless integration of gravity with quantum mechanics is something classical theories have notably struggled with.

To illustrate these concepts visually, consider a simple analogy depicted in Figure 1; think of a violin string. When played, it can appear as merely moving point-like entities producing sound. Classical physics would study these points as separate from the dynamics of the string itself. String theory, however, sees these points as manifestations of the string's vibrational modes—each note akin to a different particle.

String Vibrational Modes

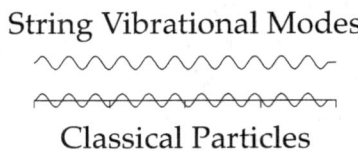

Classical Particles

Figure 1.1: Comparison of classical point-like particles versus string vibrations

This integration not only enhances our theoretical understanding but also opens up powerful avenues for predicting and explaining phenomena that otherwise remain elusive in classical terms. Through this endeavor, string theory not only reconfigures our fundamental descriptions of particles and forces but also provides a fertile framework for addressing more complex physics puzzles across the smaller and larger boundaries of the universe.

In exploring these areas where classical physics and string theory diverge, we deepen our appreciation for the intricate and beautifully interconnected fabric of the cosmos—a fabric woven not from isolated points but from dynamic, intersecting strings.

1.6 Fundamental Components: Exploring Strings

Strings, as envisioned in string theory, are fundamentally different from the particles described in classical particle physics. Rather than point-like particles, strings can be visualized as one-dimensional objects with a length but no other dimensions, akin to incredibly tiny pieces of string. Such objects can either be closed loops or open ended, each configuration contributing differently to the properties of the particles they represent. To illustrate, the visualization of these strings could be depicted using a simple line segment for an open string and a loop for a closed string, highlighting the basic nature of these elements.

In mathematical terms, the typical length scale of these strings is on the order of the Planck length, approximately 10^{-35} meters, which is many orders of magnitude smaller than what is observable with current technology. Therefore, understanding and describing their properties requires a reliance on complex mathematical frameworks such as conformal field theory and quantum mechanics.

One of the most intriguing aspects of these strings is their mode of vibration. Each string can vibrate in numerous modes, much like the varying harmonics produced by vibrating musical instrument strings. However, unlike musical instruments where each harmonic corresponds to a discrete musical note, each vibrational mode of a string corresponds to a different particle. The mass and charge of these particles are determined by the string's vibrational state, making vibrations critically important to the theory's success in describing all known particles and forces.

Consider a simple analogy with a guitar string. When plucked, a guitar string vibrates, and each mode of vibration produces a distinct sound. Similarly, each vibrational pattern of a fundamental string in string theory manifests as a particle in spacetime, with properties dependent on the vibrational characteristics. Amid this spectrum, one particular mode stands out for producing a particle with zero mass and spin-2, which in string theory is identified as the graviton, the hypothetical quantum of gravity.

The energy, tension, and types of vibration of these strings lead to a rich tapestry of theoretical predictions. The tension in a string, typically denoted as T, is another critical factor. High tension leads to stiffer strings which vibrate at higher frequencies. This energy can be expressed as $E = T \cdot L$, where L is the length of the string. Due to the extremely small scale of string length, even a minuscule amount of length results in an enormous amount of energy, tying into $E = mc^2$ to relate energy to mass and thereby influencing the mass of the particles.

Comparing theoretical predictions with experimental observations remains challenging due to the extremely high energies required to probe scales on the order of the Planck length. Current accelerator technologies can explore up to about 10^{-19} meters, indicating that direct observations of strings are beyond our reach with present capabilities. Nevertheless, indirect evidence such as the consistency of theoretical predictions with phenomena like black hole physics and early universe cosmology gives credibility to the theory.

The exploration and insights provided by theoretical studies into string vibrations offer profound implications for understanding the fundamental structure of matter and the fabric of the cosmos. Through studying how these strings vibrate, interconnected with the forces and fields acting upon them, string theory not only bridges many theoretical gaps but also opens new avenues for understanding realms beyond visible matter, such as dark matter and the very early universe.

In ensuring a full understanding of string theory's basics, it becomes evident that these fundamental strings are not merely minuscule threads vibrating in theoretical frameworks but are rather key components that may well be integral to unraveling the mysteries of the universe. As research progresses, both in theoretical calculations and experimental setups aiming at higher energies, we inch closer to actualizing the profound potential that the concept of strings holds within modern physics.

1.7 How String Theory Resolves Particle Physics Dilemmas

String theory presents an elegant solution to several persisting dilemmas in particle physics, where traditional frameworks like the Standard Model and general relativity encounter limitations. One of the foremost dilemmas is the issue of quantum gravity. The theory provides a framework that seamlessly integrates quantum mechanics with gravity, a feat that has so far eluded physicists using conventional theories.

At the core of particle physics lie fundamental particles, the smallest building blocks of nature. In standard quantum mechanics and general relativity, these particles are treated as point-like objects, which lead to significant mathematical inconsistencies, particularly when dealing with extreme scales like those found at the center of black holes or during the big bang. These inconsistencies manifest as singularities, where the laws of physics as currently understood cease to function.

Unlike point particles, strings in string theory have a finite length, albeit incredibly minute, approximately at the Planck length scale (10^{-35} meters). Each string can vibrate at different frequencies and modes, just as a guitar string vibrates in different patterns to produce various musical notes. Every mode corresponds to a different particle, with its mass and force charge determined by the string's vibration pattern. Hence, instead of countless fundamental particles, string theory posits that everything is made up of one type of string, simplifying the zoo of particles into harmonics of fundamental strings.

This approach solves several issues:

- **Ultraviolet Divergences**: The finite size of strings eliminates the problem of point-like particles that lead to ultraviolet divergences in quantum field theory. By having a non-zero size, strings do not allow the physical parameters (such as electric fields at the points constituting the particles) to become infinitely large, thus smoothing out the short-range behaviors that lead to these divergences.

- **Singularity Resolutions**: In classical general relativity, the descriptions of spacetime singularities, like those at the center of black holes or at the Big Bang, lead to incomplete theories. Strings have no definite position, diffused over a small but finite area of space, hence significantly mitigating the problems associated with singularities.

- **Integration of Forces**: Quantum field theory effectively describes three of the four fundamental forces – electromagnetism, strong nuclear force, and weak nuclear force – but struggles with gravity. String theory naturally incorporates gravity by assuming gravitons, hypothetical massless particles that mediate gravitational forces, are simply a type of vibrational mode of strings.

The implications of string theory extend beyond solving theoretical problems. For example, the idea of multiple dimensions, suggested necessary by the mathematics of string theory, offers novel ways to potentially unify all fundamental interactions into a single theoretical framework. This unification, commonly referred to as "Theory of Everything," attempts to resolve the grand conundrum of modern physics by suggesting that seemingly separate forces are manifestations of one fundamental force, described at various vibrational states of strings.

Given this background, it becomes apparent why string theory is not merely an extension of previous theories but rather a profound leap in the understanding of the fundamental principles that govern our universe. String theory redefines the conceptual frameworks upon which both quantum field theory and gravitational physics are built, offering not just solutions to old problems but also a new way of perceiving and describing the universe.

1.8 Understanding the Significance of Vibrating Strings

Throughout the discourse on string theory, one cannot overlook the pivotal role played by the simplest yet most profound element: the vibrating strings. These are not just tiny loops or strands; these are the fundamental constituents of all matter and forces in the universe according to string theory. When these strings vibrate, they do so in multiple modes and frequencies, each corresponding to a different type of elementary particle in nature. This connection is what makes studying the vibrations of strings so crucial.

The vibrations of strings can be likened to the vibrations of a violin string. Just as adjusting the tension and length of a violin string changes its resonant frequency and the sound it produces, the energy and the way in which a string vibrates determine the properties of elementary particles like electrons, quarks, and photons. The spectrum of possible vibrations is vast, suggesting a universe rich in the variety of particle types and phenomena.

A deeper look into how vibrations entail mass and force charges provide insights into the unification of laws governing the universe. Each vibrational state of a string corresponds to a particle with specific mass and force charge. These charges lead to interactions with other strings, creating the complex web of interactions we observe as forces in our macroscopic world.

The mathematical representation of vibrating strings is centered around the concept of tension (T), akin to the tension in musical strings. In string theory, this tension is enormous, which implies that the small scale of strings (approximately 10^{-35} meters) results in huge energy densities. The equations used to describe string vibrations are derived from quantum mechanics and general relativity, forming a set of complex models that allow physicists to predict new particles and the outcomes of high-energy physics experiments.

To exemplify the phenomenon in a comprehensible model, consider a visualization using TikZ:

Vibrating String

Figure 1.2: Simplified model of a vibrating string

Moving from visualization to theoretical implications, when these strings vibrate in compact extra dimensions—as proposed in various models like M-Theory—the resulting patterns are even more complex. The vibrations encompass not only the known dimensions but also those that are curled up and minute, influencing characteristics we cannot yet fully detect or measure directly.

Furthermore, the significance of these vibrating strings stretches into how they insight into the symmetries inherent in nature and how these symmetries break down. String theory beautifully demonstrates how these broken symmetries link to the creation of distinct particles and forces during the early moments of the Big Bang, proposing thus a unified approach to understanding both matter's minute scales and the vast echelons of cosmology.

In essence, the study of these vibrations unlocks potential pathways to reconciling some of the most pressing conflicts in modern physics, paving the way toward a theory of everything. They inform us not merely about the particles and forces we understand today but prospect theories on dark matter and energy, gravitational waves, the nascent conditions of our universe, and even other possible universes within the multiverse model.

1.9 String Theory's Answer to Quantum Gravity

String theory, a theoretical framework in which the point-like particles of particle physics are replaced by one-dimensional objects known as strings, provides a compelling answer to one of the most vexing problems in theoretical physics: quantum gravity. This concept refers to the theoretical foundation combining classical gravity with quantum mechanics.

The primary challenge in formulating a theory of quantum gravity lies in the reconciliation of general relativity, which describes gravitation at a macroscopic scale, with quantum mechanics, which governs the subatomic world. General relativity models gravitation as the curvature of spacetime caused by matter, but it breaks down at singularity points, like those at the centers of black holes or during the universe's inception. Here, the scales are so small and the gravitational fields so strong that quantum effects become essential. On the other hand, quantum mechanics, with its inherent probabilistic nature, fails to accommodate the deterministic laws of classical gravitation.

String theory addresses these inconsistencies by positing that the fundamental constituents of the universe are not zero-dimensional point particles, but rather tiny, vibrating strings. These strings can oscillate at various frequencies, and their vibrations correspond to different particles. The innovative part of string theory in the context of quantum gravity is how it incorporates gravitons — hypothetical quantum particles that mediate gravitational force.

Mathematical Insight into Gravitons in String Theory

In string theory, particles are depicted as excitations of strings. Photons, gluons, and even gravitons are unified within this framework. A graviton in string theory is understood as a particular vibrational state of a string. When two strings interact, the exchange of a graviton transmits the gravitational force. What makes string theory particularly suitable for quantum gravity is that these interactions naturally incorporate the quantum mechanical aspects with the geometric interpretation of gravitation.

$$S = \int d^D x \sqrt{-g}(R + \mathcal{L}_{matter})$$

Here, S represents the action integral, R is the Ricci scalar curving space and time, g corresponds to the determinant of the metric tensor, and $\mathcal{L}_{matter}$ encapsulates matter fields. In string theory, this traditional expression from general relativity is modified to include terms that describe how strings spread through spacetime and interact.

The Role of Extra Dimensions

String theory famously requires extra spatial dimensions for mathematical consistency — typically 10 or 11, depending on whether one is dealing with superstrings or M-theory, respectively. The insertion of extra dimensions allows for solving the issue where traditional quantum field theories predict infinite probabilities, an anomaly cured by the natural regularization imposed by string theory.

$$R_{MN} - \frac{1}{2}g_{MN}R = 8\pi G T_{MN}$$

In the equation above, R_{MN} is a scaled version of the Ricci tensor modified to include extra dimensions; G is Newton's gravitational constant adjusted accordingly; and T_{MN} symbolizes the stress-energy tensor incorporating contributed energies from various vibrational modes of strings.

Through the introduction of these equations and representations, string theory not only bridges our understanding of quantum physics with gravitational forces but also opens up venues to explore higher dimensional spaces that compactify into the dimensions we observe, thus maintaining consistency with observable phenomenology.

By linking graviton exchange with quantized modes of string vibrations, string theory offers a framework that is finite, predictive, and compatible within the realms of both quantum mechanics and general relativity. Furthermore, exploring these theoretical predictions paves the way for revolutionary insights into the fabric of our universe, potentially unleashing new gravitational phenomena that could be tested in future experiments or observations.

1.10 String Theory Vs. General Relativity: A Comparison

Throughout this chapter, we have navigated the intricate landscape of string theory, dotted with its rudimentary premises and profound implications. However, to fully appreciate its innovative horizon, we must juxtapose it against the venerable theory of general relativity, a cornerstone of modern physics conceptualised by Albert Einstein in the early 20th century.

General relativity revolutionized our understanding of gravity. It introduces a four-dimensional spacetime continuum where massive objects curve spacetime itself, influencing the paths that objects travel. This curving of spacetime by mass and energy is often visualized through the analogy of a heavy ball placed on a stretched rubber sheet, causing it to deform. Such a model has stood the test of time by providing extensive and accurate predictions, such as gravitational lensing and the precession of Mercury's orbit.

In contrast, string theory spans beyond the confines of general relativity by proposing that the most elementary constituents of the universe are not zero-dimensional points, but rather one-dimensional strings. These strings vibrate at specific resonant frequencies, each mode representing a different particle. The theory extends into multiple dimensions – typically 10 or 26 – beyond the perceivable four, which brings forth novel concepts such as compactified dimensions invisible to human experience.

General Relativity	String Theory
Focuses on the gravitational force	Incorporates all fundamental forces
Based on a continuum spacetime	Introduces a quantized spacetime
Operates within four dimensions	Proposes extra dimensions (usually 10 or 26)
Describes gravity via spacetime curvature caused by mass	Describes particles and interactions via vibration modes of strings
Has been confirmed by experiments like gravitational waves	Still largely theoretical with indirect supporting evidence

Despite the contrasts, one might wonder about the possible convergence of these two frameworks. String theory ambitiously seeks to provide a quantum mechanical description of gravity, something general relativity does not address. General relativity breaks down under extreme conditions, such as inside black holes or at the time of the Big Bang, where quantum effects dominate. String theory's framework, which includes tiny, vibrating strings that are never point-like, delivers finite predictions under these same extremes, resolving some of the infamous singularities predicted by general relativity.

Furthermore, integrating general relativity with string theory has led to intriguing theoretical advancements such as M-theory. M-theory proposes an 11-dimensional universe where membranes (branes) and strings are fundamental constituents. This unification under a single umbrella theory points towards a possibility of achieving the elusive theory of everything which encompasses all particles and fundamental forces of nature.

In crafting our understanding of the cosmos, general

relativity has painted a monumental picture of massive scales, cosmic events, and the flowing canvas of space-time. On the other hand, string theory offers a microscopic brush that attempts to detail every quantum weave making up the universe's fabric. While both theories may appear divergent in methodology and scale, together they complement each other—each addressing facets of our universe that the other cannot explain alone.

These theories collectively tug at the threads of our cosmic curtain, each pulling towards a grander unification that promises not only to redefine our comprehension of the universe but also to rigorously test our technologies and philosophies about the very essence of space, time, and reality. As we propel into deeper realms of theoretical physics, the interplay between string theory and general relativity continues to be a pivotal dance—one that might eventually lead to a symphony of universal truths.

CHAPTER 1. INTRODUCTION TO STRING THEORY: THE BASICS

Chapter 2

Historical Evolution of String Theory

This chapter charts the development of string theory from its inception in the 1960s to its current status in theoretical physics. It examines the pivotal moments and breakthroughs that have shaped the theory, including the identification of the five consistent string theories and the advent of M-theory. The narrative provides insights into the significant figures who have influenced the theory's trajectory and discusses the evolutionary milestones that have impacted high-energy physics, illustrating how string theory has grown from a speculative idea into a leading candidate for the theory of everything.

2.1 Early Theoretical Ideas Paving the Way for String Theory

The initial impetus for what would later evolve into string theory was rooted in a quest to understand the strong nuclear force - one of the four fundamental forces in nature, responsible for holding the nuclei of atoms together. This exploration into nuclear force interactions in the early 20th century inadvertently set the stage for the development of string theory.

In the 1940s and 1950s, theoretical physics was dominantly focused on quantum mechanics and its integration with electromagnetism, known as quantum electrodynamics (QED). However, these principles were insufficient to explain the properties of the strong nuclear force. When attempts were made to apply QED-like approaches to the strong force, calculations led to infinities that could not be realistically interpreted or removed, unlike in QED where such infinities could be renormalized.

This led physicists to explore alternative models. One significant approach was the S-matrix theory proposed by Werner Heisenberg in the late 1940s. He suggested focusing only on observable quantities like scattering amplitudes, avoiding the need to speculate about the unobservable like point-like particles and their paths. Heisenberg's idea, although initially sidelined, provided an essential conceptual framework away from field theory, emphasizing instead the mathematical consistency and symmetry in understanding physical phenomena.

Around the same time, other physicists such as Murray Gell-Mann and George Chew were independently developing the "Bootstrap model" in the late 1950s and early

2.1. EARLY THEORETICAL IDEAS PAVING THE WAY FOR STRING THEORY

1960s. This model posited that particles were not fundamental but emerged dynamically from their interactions. The self-consistency and symmetry of these interactions determined the particle spectrum. This was a radical shift from the particle-centric view of matter, suggesting instead a form of democratic "particle democracy" where all particles were treated as equally fundamental.

The Bootstrap model laid conceptual groundwork by suggesting that understanding the strong force required moving beyond traditional particle physics to consider a fully interacting system's dynamic and collective properties. However, it was inherently complex and struggled to produce precise predictions that could be tested experimentally.

Another precursor to string theory was the development of Regge theory in 1959 by Tullio Regge. Regge's work introduced the concept of Regge trajectories, which are plots relating the spin of a particle to its mass squared. These trajectories provided a new way to categorize particles based on symmetry properties, offering insights into the relationship between particle dynamics and their fundamental properties. The linear relationship observed in these trajectories hinted at an underlying order and symmetry more naturally explained later by string theory's extended objects rather than point-like particles.

The transition from these theories to string theory began with the work of Gabriele Veneziano in 1968 when he discovered that the Euler Beta function, which he was exploring for entirely different reasons, unexpectedly provided a perfect description for scattering amplitudes of hadrons. His formula encapsulated many features thought to be unique to strong interactions, such as high-energy behavior consistent with Regge theory and low-energy scattering resembling results from current algebra.

Veneziano's work did not explicitly involve strings. However, it showed that mathematical functions originally developed for number theory could successfully describe particle physics phenomena, bridging these fields in an unexpected synthesis. This serendipitous discovery prompted further investigations, which ultimately recast Veneziano's formula as the scattering amplitude for one-dimensional "strings" rather than point particles.

It is clear how theoretical ideas building from Heisenberg's S-matrix to the bootstrap hypothesis, Regge theory, and finally Veneziano's amplitude discovery, collectively navigated the course towards a string-based description of fundamental physics. The journey from empirical observation to mathematical abstraction, through attempts at theoretical consolidation under these frameworks, illustrates a rich prehistory to string theory's eventual formulation.

2.2 The First Formulation of String Theory in the 1960s

During the early 1960s, the newly formulated string theory emerged not from an attempt to unify gravity with other fundamental forces, but rather as a framework to address the complexities of hadron physics. The discovery of a plethora of hadrons in particle accelerators posed significant challenges to physicists, prompting the need for a unifying framework. The first formulation was provided by physicist Gabriele Veneziano, who stumbled upon a mathematical formula designed to explain the strong nuclear force. His hypothesis was based on the

2.2. THE FIRST FORMULATION OF STRING THEORY IN THE 1960S

Euler Beta function, which remarkably described the scattering amplitudes of hadrons.

The Euler Beta function's compelling fit with experimental data tempted Veneziano to delve deeper. This exploration led him to postulate that perhaps these hadrons were not point-like particles, but instead could be conceptualized as one-dimensional "strings". The implications of such a hypothesis were profound: if hadrons were indeed strings, the array of observed phenomena, including their resonances at certain energies, could be elegantly explained by the vibrational modes of these strings.

Veneziano's work quickly caught the attention of other physicists, including Yoichiro Nambu, Holger Bech Nielsen, and Leonard Susskind, who further refined the string concept. They introduced the idea that these strings could have different modes of vibration—quantized harmonics—that corresponded to the various particles and resonances observed in nature. Each mode was identified with a specific particle type, aligning with quantifiable properties like mass and spin.

As the theory developed throughout the 1960s, the concept expanded beyond hadrons. It was proposed that string theory could have wider applicability in particle physics. The foundational work during this period set the stage for string theory's later renaissance in the 1980s. One of the crucial advancements made by Nambu was the formulation of the Nambu-Goto action, which describes the dynamics of a relativistic string. The action is integral in quantizing string theories and is expressed as:

$$S = -T \int d\tau d\sigma \sqrt{-\det \gamma_{ab}},$$

where T represents the tension of the string, τ and σ are parameters on the worldsheet of the string, and γ_{ab} is the induced metric on this worldsheet.

This advancement illuminated the underlying mathematics of string theory, revealing that the strings' vibrations could encompass all known particles. The implications were far-reaching, suggesting a universe fundamentally composed of these vibrating strings—each harmonic potentially describing different fundamental particles.

Together, these insights from Veneziano, Nambu, and others during the 1960s laid down the conceptual underpinnings of string theory. They transformed the theoretical landscape by proposing a radical departure from point particles to strings—a shift that would eventually evolve into string theory's current prominence as a potential theory of everything. This groundwork highlighted not only new theoretical possibilities in particle physics but also indicated a path forward for the integration of gravity through further theoretical developments in subsequent decades.

2.3 The Renaissance of String Theory in the 1980s

The 1980s heralded a dramatic revival of interest in string theory, propelled by a convergence of advancements that solved critical problems and overcame the limitations confronted in the 1970s. This period, often referred to as the "first superstring revolution," was defined by a series of discoveries that shifted string theory from the periphery of theoretical physics to its center stage, promising insights into unifying the fundamental forces of nature.

2.3. THE RENAISSANCE OF STRING THEORY IN THE 1980S

The resurgence is most often credited to the groundbreaking work on anomaly cancellation by Michael Green and John Schwarz in 1984. Through their meticulous calculations, Green and Schwarz demonstrated that in ten dimensions, the type I string theory could inherently resolve the issue of anomalies, which are inconsistencies that arise within quantum field theories when attempting to gauge symmetries in the presence of gravity. Their findings, published in a landmark paper, showed that with the inclusion of supersymmetry—a theoretical framework that posits a symmetry between bosons and fermions—the anomalies could cancel out, thus preserving gauge symmetry and rendering the theory mathematically consistent.

This anomaly cancellation was critical; it not only vindicated the mathematical soundness of string theory but also hinted at string theory's capacity to underpin a unified theory of quantum gravity. The excitement generated by Green and Schwarz's work was palpable, leading to an influx of research and interest that moved string theory into the limelight.

Parallel to this, the notion of supersymmetry fortified string theory on several fronts. Theoretical physicists recognized that supersymmetric string models were more amenable to calculations and less prone to the types of quantum inconsistencies that non-supersymmetric models faced. This period also saw the increased formulation and study of heterotic string theory—a hybrid string theory that combined aspects of both closed and open strings and was crucial in compactification discussions, which addressed how extra dimensions could be mathematically reduced in physical theories.

Edward Witten, a prominent figure in theoretical physics,

contributed significantly during this epoch. His insights into the field equations of string theory and their solutions expanded the understanding of string dynamics and interactions. Witten's work led to deeper investigations into how these higher-dimensional models could physically manifest in a four-dimensional spacetime, essentially supporting the notion that all matter and force particles could emerge from the vibrations of one-dimensional strings.

To encapsulate the mathematical complexity and the broad implications of these theories, visualization through enhanced graphical representations became invaluable. Consider a two-dimensional plot illustrating the relationship between the scale of string interaction and the energy levels at which anomaly cancellations occur:

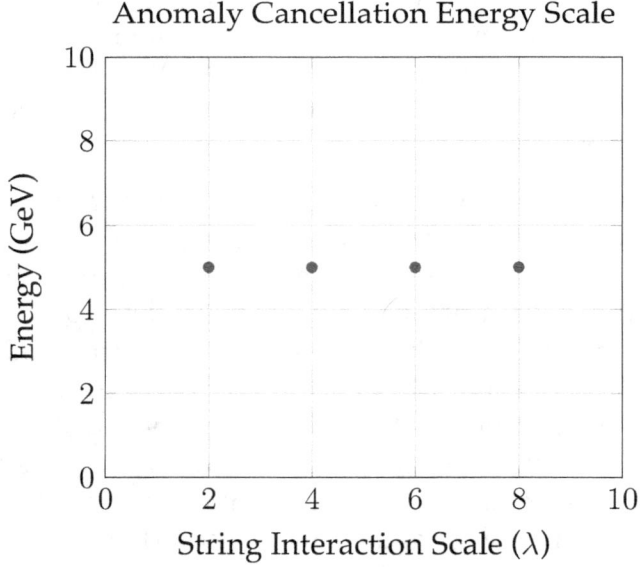

This plot indicates discrete points where anomalies cancel, showing their independence from the scale of string interactions emphasized by theoretical predictions.

In tandem with these theoretical advancements, the decade also witnessed a blossoming of collaborative efforts and workshops, where ideas were exchanged and refined. The establishment of annual string theory conferences began during this period, laying down a tradition that continues to bolster community engagement and intellectual exchange among physicists across the globe.

The implications of these advancements extended beyond mere academic interest; they provided a coherent framework that could potentially reconcile general relativity with quantum mechanics, thereby fulfilling one of physics' most profound quests. As string theory moved through the 1980s, it shaped not just a new landscape for theoretical physics but also a new vision for understanding the universe at its most fundamental level.

2.4 Development of the Five Consistent String Theories

The inception of the five consistent string theories marks a critical phase in the narrative of string theory, distinguished by a period of deep theoretical introspection and revelation. These theories, identified separately but found to be interrelated through intriguing symmetrical and dualistic properties, established the groundwork for what would later transition into the revolutionary concept of M-theory.

The five string theories emerged during the late 1980s and early 1990s, and are typically classified as: Type I, Type IIA, Type IIB, Heterotic $SO(32)$, and Heterotic $E_8 \times E_8$.

Each of these theories operates under the fundamental premise of superstring theory, which suggests that elementary particles are not point-like dots but rather oscillating strings. The vibrational mode of the string determines the type of particle rendered, affording a variety of phenomenological outcomes that correspond with known particle behaviors.

The development of Type I string theory introduced open strings with endpoints that can attach to D-branes, alongside closed strings. This theory includes both gauge and gravitational interactions but demands the existence of unoriented strings, implying that some physical processes described by Type I strings are indistinguishable when time or spatial orientations are reversed.

Type IIA and Type IIB string theories expanded on the concept of closed strings alone but diverge in their treatment of supersymmetry. Type IIA theory preserves supersymmetry in a non-chiral fashion, meaning it includes both left- and right-handed supersymmetric partner transformations equally. In contrast, Type IIB theory endorses chiral, that is direction-oriented, supersymmetry which differentiates its mathematical and physical implications.

Both heterotic theories, Heterotic $SO(32)$ and $E_8 \times E_8$, interestingly combine the ten-dimensional superstrings with a 26-dimensional bosonic string model, compactified on a 16-dimensional torus to reconcile the manifold dimensions. This intricate combination, supported by anomaly cancellation that essentially keeps theoretical consistency, accommodates gauge interactions in a unified string framework.

These theories were initially considered independent and discrete frameworks for understanding string dynamics.

However, revelations emerging from dualities amongst them pointed to a profound interconnectedness. Dualities are transformation abilities that allow one to convert entities in one theoretical framework into their counterparts in another, without loss of physical properties. These include T-duality and S-duality, which respectively relate string theories in compactified dimensions through radius inversion and coupling constants inversion.

For visual support, consider the tikz illustration below to understand the primary connections via dualities between these theories:

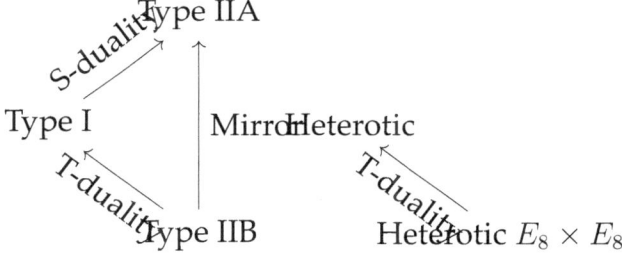

As these connections suggest, the discovery of dualities served as both a unifying and simplifying principle, steering the research community toward the hypothesis that all five string theories could be mere reflections or facets of a more profound singular framework—an idea that serves as an introductory path to M-theory.

2.5 Introduction of D-branes in the 1990s

The 1990s witnessed a seismic shift in the landscape of string theory with the introduction of D-branes, a concept that profoundly extended our understanding of the theory's underlying structure and its applications. Introduced by Joseph Polchinski in 1995, D-branes—short for Dirichlet boundary conditions—represent locations at which strings, the fundamental elements of string theory, can end or be constrained. This was a significant diversion from the prior belief that strings could only be closed loops, without ends.

D-branes are more than mere endpoints for strings; they are dynamical objects with their own physical properties, such as mass and charge. The realization that these branes could carry such attributes brought about a new layer of complexity and richness to string theory, tying it more closely to other areas of physics, particularly gauge theories and quantum field theories.

One of the key breakthroughs brought about by D-branes was their role in realizing that objects in string theory can have multiple dimensions. While traditional string perturbation theory treated strings as one-dimensional objects, D-branes introduced the possibility of higher-dimensional objects within string theory. This ranged from one-dimensional lines (which would be a string itself) to surfaces (two-dimensional planes) and higher-dimensional analogs.

The mathematical formalism underlying D-branes fundamentally shifted with Polchinski's demonstration that these objects could be described in the language of differential geometry and gauge theory. In practical terms, this relates to how D-branes interact with the fields in string theory; they can host what are known as gauge fields, crucial components in field theories that describe the fundamental forces of nature.

The implementation of D-branes into string theory also catalyzed new subfields of research within the broader framework of theoretical physics. A particularly influential development was the realization of the gauge/gravity duality, also known as the AdS/CFT correspondence, proposed by Juan Maldacena in 1997. This duality suggested a profound connection between certain gauge theories (describing quantum fields) and theories of gravity (describing spacetime and gravitational interactions), mediated through the behavior of D-branes.

To elucidate the significance of D-branes in the broader context of theoretical physics, consider their impact on the study of black holes. The microscopic properties of black holes, which were for decades a mystery, began to be understood by analyzing the dynamics of D-branes. This has been crucially important for exploring quantum aspects of gravity and for potential resolutions to paradoxes such as black hole information loss.

The aftermath of Polchinski's introduction of D-branes has been characterized by a profusion of research leading to new conjectures, theories, and mathematical tools evolving from or related to D-brane dynamics. These include various non-perturbative techniques in string theory, implications for cosmology and particle physics, and a deeper understanding of the fundamental structure of matter and spacetime.

As we delve further into the complexities and applications of D-branes, we continue to unravel more about the universe at both cosmic scales and at the scale of fundamental particles, with D-branes acting as a crucial bridge between theoretical predictions and the empirical reality we observe in high-energy physics experiments. The introduction of D-branes not only broadened the scope of string theory but also anchored it more firmly as a pivotal element in the quest for a unified theory of every force and particle in the universe. Their ongoing impact suggests that they will remain at the heart of theoretical physics discussions and developments for years to come.

2.6 The Second Superstring Revolution and Dualities

The period termed the Second Superstring Revolution, occurring predominantly during the mid-1990s, marked a profound transformation in the field of string theory. Central to this revolution were the discoveries of new symmetries and dualities within the existing frameworks of string theory, which significantly enhanced the understanding and connections between these frameworks.

Dualities, in the context of string theory, refer to surprising mathematical equivalences that relate seemingly different theories. These dualities allow physicists to translate difficult problems in one theory into simpler problems in a dual theory, acting as a bridge across theoretical landscapes. There are primarily three types of dualities identified: *S-duality*, *T-duality*, and *U-duality*.

S-duality reveals a relationship between string theories with strong coupling constants and those with weak constants. This duality suggested a promising perspective by showing that calculations which are intractable in a strong coupling regime of one theory could be more manageable in the weak coupling regime of a dual theory. For instance, Type IIB string theory is S-dual to itself–the strongly coupled Type IIB string theory can be described by translating it into the weakly coupled domain of the same theory.

2.6. THE SECOND SUPERSTRING REVOLUTION AND DUALITIES

T-duality emerged as an astonishing aspect of string theories where the compactification or the wrapping of strings around small circular dimensions led to actionable mathematical correspondences. A quintessential manifestation of T-duality is between Type IIA and Type IIB theories; when strings are compactified on a circle, shifting from a smaller to a larger radius in one theory (Type IIA) translates to details about a large to small radius in the other (Type IIB) and vice versa.

U-duality, introduced by Cremmer, Julia, and Lu in their study of eleven-dimensional supergravity (which plays a pivotal role in M-theory), is a synergy of both T- and S-dualities offering a unified framework through which various string theories could be inter-connected. This not only consolidated the relation among different string theories but also reinforced the enduring quest for a singular unifying theory, M-theory.

M-theory emerged from these insights, as discussed earlier in this chapter—a theory that encompasses all five previously distinct string theories into a single framework by integrating them through various dualities. Edward Witten was a major proponent of this revolutionary idea, presented famously in his discourse at the String 1995 conference, calling this unifying theory M-theory where 'M' implies 'magic,' 'mystery,' or 'membrane' according to different interpretations proposed by Witten himself.

Integral to these theoretical advancements were the contributions involving higher-dimensional objects known as **D-branes**. These are dynamic objects where open strings can end, and they occupy pivotal roles in understanding the non-perturbative aspects of string theories (especially in Type I and Type II strings) and are essential for the realization and evidence of dualities. D-branes provided concrete entities that could be used to probe various properties of the string theories including black hole physics, strongly coupling symmetries, and gauge-field interactions.

The discovery and exploration of these dualities during the Second Superstring Revolution thereby nudged string theory towards an even more potentially verifiable scientific framework than before, suggesting ways to unify the elemental forces of nature in dimensions and realms previously unimagined. These advances inherently beckon further scrutiny and testing, wherein lies the potential for unveiling more about our cosmos—its origins, its most foundational layers, and perhaps clues to its ultimate fate.

2.7 The Era of M-theory and the Theory of Everything

The unification of the different string theories into a single framework known as M-theory marked a significant turning point in the historical development of string theory, acting as both a culmination and a fresh departure point within the field. Emerging in the mid-1990s following a series of conjectures and insights primarily by physicists Edward Witten and Paul Townsend, M-theory proposed a radical idea that the five different string theories are not distinct entities but different limits of a single underlying theory.

This proposal was groundbreaking because it suggested an eleven-dimensional universe, where the additional dimension opened up new possibilities for solving previ-

2.7. THE ERA OF M-THEORY AND THE THEORY OF EVERYTHING

ously intractable problems in physics. The presence of this extra dimension, as well as the theoretical entities known as membranes, or branes, brought about a shift in focus for string research, from traditional string-like one-dimensional objects to higher-dimensional objects which could include two-dimensional membranes and other p-dimensional entities generally referred to as p-branes.

Dimension	Theoretical Entity	Role in M-theory
1D	Strings	Fundamental building blocks
2D	Membranes	Stabilization of extra dimensions
pD	p-branes	Generalized objects in higher dimensions

M-theory also sparked enhanced interest in supergravity theories, particularly in eleven dimensions, which were now understood to be part of the broader M-theory landscape. These developments necessitated new mathematical tools and approaches, leading to advances in fields such as algebraic geometry and differential topology to handle complex calculations involving multidimensional spaces.

One of the landmark theoretical developments enabled by M-theory was the greater understanding of black holes. The theory provided a framework in which the entropy and temperature of certain classes of black holes could be calculated exactly, through the study of their microstates in string theory terms. This achievement was not only a significant test of string theory's validity but also provided profound insights into aspects of quantum gravity.

The narrative of unification encapsulated by M-theory brought with it new challenges and opportunities. The theory required physicists to ponder not just the strings themselves but also their interactions and the dynamics of spacetime at scales much tinier than those previously considered. This paradigm shift brought string theory into closer contact with experimental possibilities, potentially observable through indirect cosmological effects or future particle physics experiments.

M-theory's introduction into the mainstream of theoretical physics serves as a pivotal moment, indicating not just a technical development but also a conceptual evolution. It promises a unification not merely of string theories but hints at a deeper structure underlying all physical laws. As research continues, M-theory remains a central topic in theoretical physics, with implications that extend into cosmology, quantum mechanics, and beyond, continually driving forward our understanding of the Universe's most fundamental workings.

2.8 Impact of String Theory on High-energy Physics

The incursion of string theory into the realm of high-energy physics has been profound and multifaceted, influencing both theoretical perspectives and experimental methodologies used in particle physics. To understand this influence comprehensively, we must look at key areas such as the unification of forces and particles, quantum gravity effects, and the implications for early universe cosmology.

One of the primary impacts of string theory is its role in proposing a unified framework for all fundamental forces and types of matter in the universe. Unlike the Standard Model of particle physics, which exceptionally describes three out of four known fundamental forces excluding gravity, string theory ambitiously attempts to incorporate gravity by modeling particles not as zero-dimensional points but as one-dimensional strings. These strings' vibrations correspond to different particles, where the mode of vibration determines the type of particle represented. This innovative approach offers a sweeping overview of particle interactions at energy scales much beyond those currently accessible by experiments.

Moreover, string theory's introduction of additional spatial dimensions than the four (three space and one time) encountered in everyday experiences is a radical yet potentially transformative aspect within theoretical physics. These extra dimensions, compactified in intricate Calabi-Yau manifolds, provide a novel context in which gravity and particle dynamics can be explored. The mathematical tools developed for studying these manifolds have enriched the field by offering new methods and insights into geometry's role in theoretical physics.

In exploring quantum gravity, string theory has enabled significant progress. Traditional techniques used in quantum field theory are ineffective in dealing with the singularities and infinities arising at the quantum level. However, string theory provides a natural regularization mechanism due to the finite size of strings. These extended objects do not probe spacetime at infinitely small points, thus softening potential singularities seen in general relativity and offering a glimpse at the quantum structure of spacetime.

The potential insights into early universe cosmology provided by string theory have ignited considerable interest. For instance, the theory predicts a multitude of possibilities during the universe's inflationary period leading to a landscape of possible universes within a multiverse framework. These theories are intrinsically linked with observational cosmology, aiming to potentially provide empirical evidence through anomalies in background cosmic radiation or distribution of galaxies.

Additionally, the theoretical tools and concepts developed through string theory have found applications in other areas of physics. The AdS/CFT correspondence, which relates a type of string theory in anti-de Sitter space to a conformal field theory on the boundary of the space, is one such breakthrough. This duality has not only deepened our understanding of quantum gravity but has also been used as a tool in studying quantum chromodynamics, the theory of the strong force, thus influencing the way physicists calculate the properties of quark-gluon plasma.

In summary, while direct experimental evidence for string theory remains elusive, its impact on high-energy physics is undeniable. Offering innovative solutions to theoretical problems, enriching mathematical frameworks, and potentially revolutionizing our understanding of universe's properties from its smallest to largest scales, string theory continues to drive significant advances in high-energy physics.

2.9 String Theory in the 21st Century: Enhancements and Impact

The turn of the millennium marked a significant phase in the evolution of string theory, witnessing both substantial theoretical advancements and increased integration into broader physics dialogues. This period is characterized by the expansion of string theory beyond its traditional confines, interfacing with cosmology, particle physics, and even pure mathematics.

As we venture deeper into the 21st century, a pivotal focus has been on the phenomenological implications of string theory. The quest to derive testable predictions that could be feasibly verified in particle accelerators like the Large Hadron Collider (LHC) has intensified. String theory's ability to potentially explain the fine structure constant or the masses of fundamental particles continues to tantalize physicists. Although direct evidence eludes the current technological reach, indirect consequences, such as insights into the behaviors of black holes and early universe conditions, are profound.

Another noteworthy development in this era is the application of string theory to the mysteries of dark matter and dark energy. These constitute the majority of the universe's mass-energy content, yet remain poorly understood. String theory offers models that could unify these hidden components of the universe with the visible matter, mediated through extra dimensions and braneworld scenarios. Computational tools and techniques such as lattice string theory have begun to provide frameworks for simulating scenarios where these theoretical constructs might manifest.

On the technical front, the refinement of AdS/CFT correspondence—an idea posited by Juan Maldacena in late 1997—has flourished in the 21st century. This principle asserts a duality between theories of gravity on Anti-de Sitter (AdS) space and Conformal Field Theories (CFT) in one fewer dimension. It has provided powerful theoretical tools for studying quantum gravity and strongly coupled quantum field theories, offering insights into quark-gluon plasma and various condensed matter systems. The capacity of AdS/CFT to bridge disparate areas of physics is not just theoretical but has spurred multiple interdisciplinary studies linking string theory with practical experiments.

The mathematical underpinnings of string theory also saw significant strides with the development of new geometric frameworks such as derived category theory and non-commutative geometry. These advancements not only enhance our understanding of string theory itself but also enrich the broader field of mathematics by introducing new concepts and techniques.

In addition to theoretical progress, educational and public outreach about string theory has expanded. Enhanced by digital technology, open courses, webinars, and interactive textbooks have brought the high-level concepts of string theory to a broader audience. This democratization of knowledge has played a crucial role in inspiring the next generation of theoretical physicists.

As this new century progresses, the impact of string theory continues to permeate various domains, presenting a richer tapestry of interaction between theoretical insights and practical applications. The ongoing developments promise not only to answer some of the most fundamental questions about our universe but also to pose new ones, possibly reshaping our understanding of the cosmos in profound ways.

2.10 Major Figures in the Development of String Theory

The landscape of string theory, while rich in mathematical constructs and theoretical propositions, is also densely populated with brilliant minds whose contributions have profoundly shaped its evolution. To understand the advancement of string theory, it is essential to explore the seminal roles played by key physicists whose work not only propelled the theory forward but also intertwined their legacies with the very fabric of the universe as described by string theory.

Gabriele Veneziano is often heralded as a foundational figure in the field. His formulation of the Veneziano amplitude in the late 1960s provided an unexpected bridge between nuclear forces and string concepts, initially intended to describe strong nuclear interactions. Veneziano's work sparked considerable interest, laying the groundwork for further exploration into what was then not yet called "string theory."

As string theory began to gain traction, **Yoichiro Nambu**, **Holger Bech Nielsen**, and **Leonard Susskind** independently realized in the early 1970s that the Veneziano model could be explained by a physical model of strings. Their collective insight that particles could be viewed as different excitations of a fundamental string was revolutionary, providing a concrete physical picture that underpinned early string theory.

The field, however, truly began to mature as a unifying theory in physics with the groundbreaking work of **John Schwarz** and **Michael Green** during what is now fondly referred to as the "first superstring revolution" in the mid-1980s. Their discovery of anomaly cancellation in type I string theory if certain symmetries (Green-Schwarz mechanism) were present pointed to the possibility of unifying gravity with other fundamental forces, thus propelling the entire field of string theory into the limelight as a candidate for a Theory of Everything.

Contemporaneous to these developments was **Edward Witten**, whose prodigious contributions to both the mathematics and physics of string theory cannot be overstated. Witten's work in the late 1980s and 1990s helped in advancing the understanding of various dualities within string theory, which suggest that different versions of the theory are actually equivalent. His insights deepened the connection between string theory and quantum gravity and played a pivotal role during the second superstring revolution.

In understanding branes within string theory, **Joseph Polchinski**'s introduction of D-branes as tangible objects within string theory not only solved numerous puzzles but also opened up myriad pathways to link string theory with observable quantum phenomena and black hole physics. This marked a significant shift in how physicists approached string theory, integrating it more closely with experimental data.

With the advent of M-theory, a compelling unification of all five consistent string theories emerged, largely influenced by Witten's further revolutionary work and thorough explorative studies by physicists like **Paul Townsend** and **Petr Hořava**, who extended the frame-

2.10. MAJOR FIGURES IN THE DEVELOPMENT OF STRING THEORY

work of string theory beyond its original confines. Their contributions helped elucidate various aspects of eleven-dimensional theories that M-theory posited, further broadening the theoretical landscape.

These theorists, among others, have not only contributed to the technical corpus of string theory but have also dramatically enhanced its philosophical and conceptual foundations, fostering a richer dialogue between mathematics and physics. The interplay between these intellectual domains has not only enriched the academic landscape but has also fundamentally altered our understanding of the universe's architecture.

In reflecting on these contributions, it becomes evident that string theory is as much a product of human curiosity and intellectual rigor as it is a framework of the cosmos. It compels us to ponder not just the nature of reality, but also the processes through which we come to understand such profound truths about our universe. Perhaps it is in this reflective consideration where we realize that our journey into understanding fundamental physics is as intricately woven as string theory itself suggests the fabric of the cosmos to be.

CHAPTER 2. HISTORICAL EVOLUTION OF STRING THEORY

Chapter 3

Fundamental Concepts: Strings, Branes, and Extra Dimensions

This chapter delves into the primary entities and concepts at the heart of string theory, namely strings, branes, and extra dimensions. It discusses the characteristics and types of strings, introduces the concept of branes as multi-dimensional spaces within the theory, and explains the role and nature of additional dimensions beyond the familiar three spatial dimensions. The chapter further explores how these elements interact within string theory, providing a basis for much of its theoretical structure and predictions about the universe's fundamental properties.

3.1 Defining Strings: The Basic Building Blocks

Strings are hypothesized one-dimensional entities, infinitesimally small, which constitute the most fundamental constituents of the universe in string theory. Unlike point particles of traditional particle physics, strings have length but no other discernible dimensions, which sets the stage for a new paradigm in the fabric of physics. Their physical length is typically characterized by the Planck length, approximately 1.6×10^{-35} meters. While this dimension is subatomic and unfathomably small, the implications of strings' extended nature are vast and profound.

$$\text{Planck Length (L)} = \sqrt{\frac{\hbar G}{c^3}} \approx 1.6 \times 10^{-35} \text{ meters}$$

where $\hbar$ represents the reduced Planck's constant, G is the gravitational constant, and c is the speed of light. Understanding this scale is crucial because it underpins one of the foundational assumptions of string theory — that every particle traditionally perceived as a point is instead a vibrating string.

Each string can vibrate in multiple modes, and it is the vibration of a string, not its substructure, which determines the type of particle it represents. This idea introduces an elegant unifying concept: all particles are manifestations of one fundamental object, differing only in their vibrational states. This can be likened to musical instruments where different notes are produced depending on how strings vibrate.

3.1. DEFINING STRINGS: THE BASIC BUILDING BLOCKS

$$f = \frac{1}{2L}\sqrt{\frac{T}{\mu}}$$

The above formula describes the fundamental frequency f of a vibrating string where L is its length, T its tension, and μ its mass per unit length. In string theory, tension is an especially significant factor because it is tremendous, much higher than anything achievable with current materials science, and is a crucial determinant in the resulting mass-energy of particles, according to Einstein's mass-energy equivalence $E = mc^2$.

Interactions between strings, fundamental as they are, also replace the classical notion of particles interacting at points. Two strings approaching each other can merge or split, corresponding to the familiar physical processes whereby particles are either created or destroyed. These interactions imply that strings can connect at points along their length (in the case of open strings) or at their ends if they are closed loops, leading to a rich variety of possible outcomes - a true choreography at nature's most fundamental scale.

The transition from representing particles as points to modeling them as extended objects enables a divergence from past singularities like those found in black holes or at the Big Bang. String theory proposes that these singularities do not exist in the classical sense, replaced instead by finite, one-dimensional strings. Thus, while extremely tiny, strings offer a powerful and intriguing framework to approach and understand some of the most perplexing questions in physics regarding the nature of space, time, and the universe.

This shift from point particles to strings isn't merely theoretical but opens new pathways for discussing gravity and quantum mechanics within a single framework — a key quest in theoretical physics.

3.2 Types of Strings: Open and Closed Strings

In the unification attempts of fundamental forces under string theory, distinguishing between types of strings—open and closed—is pivotal. Each class exhibits unique properties and plays specific roles in the universe's fabric.

Open Strings: An open string is analagous to a line segment; it has two distinct endpoints. These endpoints can attach to other objects in the spacetime, specifically D-branes—a concept later expounded in this chapter. The vibrational modes of open strings are crucial—they do not merely vibrate in place but can exhibit more complex interactions due to their endpoints moving along the attached branes. The significance of open strings extends to their role in mediating the forces akin to electromagnetic interactions in quantum field theories.

Mathematically, the dynamics of open strings are described by letting the endpoints satisfy Neumann or Dirichlet boundary conditions, which relate physically to whether the endpoints can move freely or are fixed, respectively. The boundary conditions reflect the nature of interaction with other physical entities like branes.

Closed Strings: In contrast, a closed string forms a continuous loop without any endpoints. The absence of termination points gives closed strings symmetric properties and allows them to be more stable and less interactive

3.2. TYPES OF STRINGS: OPEN AND CLOSED STRINGS

with other types of physical entities compared to open strings. Closed strings are primarily associated with the gravitational force within string theory. This is because their vibrational modes include a graviton mode—an essential element for theories attempting to unify gravity with other fundamental forces.

In theoretical terms, the representation of closed strings involves summing up not only path integrals over all possible locations in spacetime but also integrating over all possible shapes and sizes of loops. This makes their mathematical treatment more complex and contributes to rich implications such as the graviton's emergence from string theory's framework.

Interconnecting Dynamics: Open and closed strings are two faces of the same coin. Interestingly, in certain theoretical contexts, open strings can join ends to form closed strings, and similarly, closed strings can split into open strings under appropriate conditions. This duality is fundamental in understanding scattering amplitudes and interactions at quantum levels, offering a robust framework for exploring particle physics and beyond.

This interplay and the transformation between open and closed strings are meticulous areas of study in string theory, hinting at underlying symmetries and conservation laws in intricate quantum mechanical systems. The theories suggest that these transformations are not mere theoretical constructs but have profound implications, possibly explaining fundamental forces and matter constituents in our universe.

By dissecting the properties and interactions of open and closed strings, we advance our understanding of the universe. Their study not only demystifies aspects of quantum mechanics but also serves as a bridge in the quest for a Theory of Everything that seamlessly merges the theories of large-scale gravitation with quantum mechanics. Such exploration, while challenging, continues to push the boundaries of our understanding in theoretical physics.

3.3 Introduction to Branes: Multi-dimensional Spaces

Branes play a pivotal role in string theory, serving as foundational elements that extend our understanding of the universe's basic structure. Briefly put, a brane in string theory is a physical object that generalizes the notion of a point particle to higher dimensions. Derived from the word "membrane," branes can exist in multiple dimensions. Their classification depends on the number of dimensions they span, known as 'p' in 'p-branes', where 'p' represents the spatial dimensions of the brane itself.

In simpler terms, a point particle can be seen as a 0-brane, a string as a 1-brane, and a membrane as a 2-brane. As this notion extends, we encounter higher dimensional branes like 3-branes, 4-branes, and so forth, all the way potentially up to 9-branes in theories with extra spatial dimensions. This concept significantly shifts the paradigm from one-dimensional strings to multi-dimensional entities that significantly enrich the theoretical landscape of string theory.

One crucial aspect of understanding branes is realizing that they can carry various types of physical fields on their worldvolume - the multidimensional analog of a particle's worldline. These fields include gravity, gauge theories, or matter fields. Crucially, the dynamics of branes are governed by the Dirac-Born-Infeld (DBI) action, generalized from electromagnetism, exhibiting fascinating parallels and distinctions from the much simpler action of a string described by the Nambu-Goto action.

Moreover, the introduction of D-branes (where "D" stands for Dirichlet) marked a significant development in string theory. These are branes where open strings can end, with their endpoints attached to the D-brane. Before this insight, open strings in string theory were considered to have free ends that could lie anywhere in space. The D-brane provided not only a boundary condition for open strings but also became central to understanding strong gauge dynamics due to its intrinsic connection to gauge symmetries and gauge fields.

The theories involving D-branes have also led to advancements in understanding dualities within string theory. Dualities are symmetries that allow physicists to relate one strong coupling problem to a weak coupling counterpart in another string theory. D-branes play an essential role in these dualities and thereby enrich the understanding of quantum gravity.

To visualize the role of branes within string theory, consider their function in scenarios like brane-world cosmology. In these models, our observable universe is conceptualized as a 3+1-dimensional brane embedded in a higher-dimensional space. Extra dimensions remain invisible because the forces of nature, except for gravity, are confined to the brane. Gravity's ability to leak into the bulk—the space surrounding the branes—could potentially be observable through phenomena like graviton-induced interactions, offering insights into dimensions beyond our sensory perception.

Understanding these multi-dimensional spaces requires

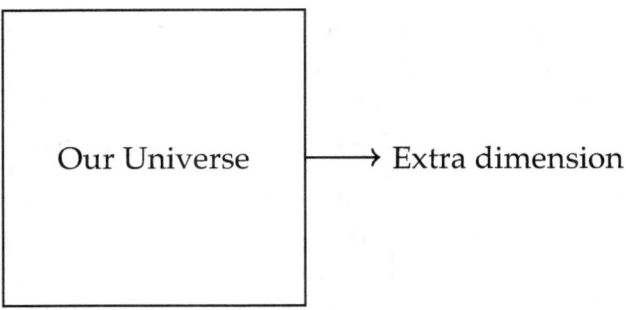

Figure 3.1: Illustration of our universe as a 3-brane embedded in a higher-dimensional space.

acknowledging that branes are not just theoretical constructs but are tools that potentially offer explanations about the universe's fabric that go beyond what point particles or strings alone have provided. Theoretical investigations into these areas continue to push the boundaries of physics, intertwining mathematical elegance with profound insights into the cosmos.

3.4 Different Types of Branes: From D-branes to M-branes

As we journey deeper into the complex topology of string theory, it is crucial to differentiate between the various classifications of branes, particularly D-branes and M-branes, each playing pivotal roles in the framework of the theory.

D-branes, or Dirichlet branes, are among the most discussed entities in string theory. These are dynamical objects to which open strings can attach their endpoints under the presence of Dirichlet boundary conditions - condi-

3.4. DIFFERENT TYPES OF BRANES: FROM D-BRANES TO M-BRANES

tions that fix the endpoints of the strings to specific points in space. Unlike strings, which are fundamentally one-dimensional, D-branes extend in multiple dimensions. The 'D' in D-brane refers to these Dirichlet boundary conditions. Mathematically, a Dp-brane is a p-dimensional object meaning that it extends in p dimensions; thus, a D1-brane is a one-dimensional line, a D2-brane is a two-dimensional surface, and so on, up to D9-branes in ten-dimensional space-time in types IIA and IIB string theories.

The duality properties within string theory often relate different types of D-branes. For instance, a type IIA theory can, under certain dualities, transform its Dp-branes into D(p±1)-branes in type IIB theory and vice versa. This feature is crucial for understanding the correspondence and possible transformations between various string theories and is one of the cornerstones of the unifying M-theory.

M-branes are integral elements in the broader scope of M-theory, which extends and encompasses all five superstring theories through 11-dimensional supergravity. There are primarily two types of M-branes: M2-branes (also known as membranes) and M5-branes. M2-branes are two-dimensional, allowing them to support a magnetic charge in the supergravity theory. On the other hand, M5-branes are five-dimensional, supporting a corresponding magnetic dual. Both types of M-branes are vital in understanding the full implications and the expansive multidimensional landscape that M-theory proposes.

To demonstrate how these branes manifest within higher dimensions, let us consider their representation in a simplified model. If we visualize a scenario where several

M2-branes and M5-branes coexist, their dynamics and interactions often result in complex geometries and configurations. These interactions lead to a variety of solutions in the 11-dimensional supergravity that support the diverse phenomena predicted by M-theory. Each type of brane contributes uniquely to the fabric of the universe as described by string theories, affecting everything from the fundamental forces identified in physics to the possible explanation of the enigmatic dark matter and dark energy.

By introducing tools from differential geometry and topology, we deepen our understanding of how branes curve space-time and influence physical processes at the most fundamental level. Techniques such as Betti numbers and homology groups provide powerful methods for classifying the topology of branes, predicting their effects on the physical universe.

Furthermore, the importance of these branes extends into real-world applications where theories incorporating quantum gravity - like string theory's implications on cosmology and black hole physics - profoundly suggest that high-energy experiments might one day detect signatures of brane configurations.

Thus, this exploration of branes from fundamental types like D-branes and M-branes discloses a more intricate view of how multidimensional entities contribute to our understanding of the universe from micro to macro scales. This context not only enriches our comprehension but also tantalizes with possibilities of new physics and applications yet to be discovered.

3.5 Concept of Extra Dimensions: Beyond the Familiar Three

The notion of extra dimensions is pivotal in understanding the complexities and the groundbreaking perspectives offered by string theory. This concept extends beyond our conventional understanding of three spatial dimensions—length, width, and height—which are accessible and observable in our everyday experiences. String theory posits that there are additional spatial dimensions, which, although not perceptible to our daily observations, play critical roles in the fundamental workings of the universe.

To begin delving into the reasons behind the proposal of extra dimensions, one must consider the mathematical formulations and physical insights required by string theory. The standard model of particle physics, which describes three spatial dimensions, is inadequate for integrating gravity with other fundamental forces—electromagnetism, weak nuclear force, and strong nuclear force. String theory, by introducing additional dimensions, offers a framework where these forces can be unified at a quantum level.

The mathematical beauty of string theory suggests that for a consistent theoretical framework, there must be more than the observable three dimensions. Early versions of the theory required the existence of up to 26 dimensions to resolve various mathematical anomalies. However, with the advent of superstring theory, which incorporated supersymmetry, this requirement was refined down to 10 dimensions—nine spatial and one temporal.

The visualization of these extra dimensions can be approached by considering them as being 'compactified', a process whereby additional dimensions are curled up so tightly that they become too small to detect at low energies or with current technological capabilities. A common analogy used to explain this concept is to consider a garden hose. From a distance, a hose appears as a one-dimensional line. Only upon closer inspection does its two-dimensional surface become apparent. Similarly, these extra dimensions are theorized to be compact and interwoven within the fabric of the familiar three dimensions at scales near the Planck length (10^{-35} meters), which is considerably smaller than can be currently probed by any experimental equipment.

The types and shapes of these compact dimensions significantly influence the properties of strings vibrating within them. The specific mode of vibration of a string determines the type of particle it represents, with different vibrational patterns representing different fundamental particles. This dependency implies a deep connection between these extra spatial dimensions and the observable properties of matter, potentially providing answers to questions about particle masses, charges, and other characteristics.

Moreover, the geometry and topology of these extra dimensions lead to rich structural features potentially observable via phenomena such as Kaluza-Klein excitations. These excitations could impart unique signatures observable at high-energy scales achievable in particle accelerators like the Large Hadron Collider (LHC). Thus, even though these dimensions are not directly observable, their indirect effects could manifest through novel physical phenomena.

The implications of extra dimensions extend beyond mere theoretical curiosity. They provide a more complete understanding of the universe's structure at its most fundamental level, offering possible explanations for the enigmatic aspects of particle physics, cosmology, and gravitation. In navigating through these intricate concepts, string theory paints a picture of a universe far more exotic and complex than the one apparent to our senses. As researchers delve deeper into the underlying structure of space and time facilitated by string theory, they continually recalibrate our understanding of the cosmos—an endeavor that spirals into ever more profound and subtle realms of reality.

3.6 Compactification: How Extra Dimensions are Hidden

To comprehend the mechanisms by which extra dimensions are concealed in string theory, we delve into the concept of compactification. Compactification is a pivotal mathematical and physical process that explains how higher dimensions can exist, yet remain perceptible only at minuscule scales, typically at the Planck length (approximately 10^{-35} meters). The process involves the dimensions being 'curled up' or compacted to such a scale that they evade detection by current experimental means.

Let us imagine, metaphorically, a garden hose lying across a vast field. From a distant vantage point, the hose appears to be a one-dimensional line. Only when one comes significantly closer does the second dimension, the circumference of the hose, become apparent. This is an analogy often used to understand how extra dimensions are concealed in the universe. These extra spatial dimensions are theorized to be tightly compacted into complex geometrical shapes, known as Calabi-Yau manifolds, which only unravel and become noticeable at incredibly small scales.

Given that string theory posits the existence of multiple extra dimensions, the role of compactification is not just to hide these dimensions but also to shape the physical constants and forces we observe in our four-dimensional universe. The specific manner in which these dimensions are compacted determines properties such as charge, mass, and force strengths. The intricate topology of Calabi-Yau spaces, characterized by holes and complex folding patterns, supports a diverse spectrum of vibration modes for the strings moving through them.

Elucidating the mathematics, Calabi-Yau manifolds are defined as n-dimensional complex manifolds that are Kähler and possess a vanishing first Chern class. These conditions ensure that they provide a rich structure required to support phenomena like supersymmetry, crucial for string theory's consistency. For visual illustration of these concepts, Figure 3.2 shows a rendered image of a two-dimensional representation of a Calabi-Yau manifold.

As physicists continue their investigations into the underlying principles of string theory, the compactification of extra dimensions remains a fertile ground of research. The mysteries nested within these minute spaces hold the answers to some of the most profound questions about the nature of our universe, such as the unification of forces and the quantization of gravity. By studying the properties and interactions facilitated by these hidden dimensions, scientists are peering deeper into the fabric

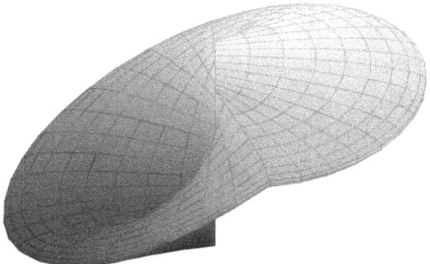

Figure 3.2: Stylized 2D representation of a Calabi-Yau manifold.

that constitutes everything in existence—the tightly woven, intricate, and beautifully complex tapestry of the cosmos. Through the progressive unravelling of these compact dimensions, future research may reveal new physics that underlines our very reality. Unseen, these dimensions offer more than just a haven for theoretical escape; they hold tangible clues poised to revolutionize our understanding in palpable ways.

3.7 The Role of Strings and Branes in Higher Dimensions

In string theory, the interaction between strings, branes, and higher dimensions forms a sophisticated symphony that shapes our understanding of the universe's fundamental fabric. Strings, the one-dimensional fundamental entities, oscillate and move in a space immersed with numerous extra dimensions, influencing the dynamics and properties of these higher dimensions.

To unpack the role of strings in higher dimensions, consider a string vibrating in a ten-dimensional space, as is typical in superstring theories. Each vibration mode of the string corresponds to a different particle in our perceivable four-dimensional universe. Thus, rather than merely existing in these higher dimensions, strings interact with them through their vibrational patterns, which encode information about particle types and their interactions.

While each string vibration can theoretically produce an infinite array of particles, the presence of extra dimensions limits these possibilities through the mechanism known as compactification. Compactification refers to the process where extra dimensions are curled up so tightly that they have a profound impact on the vibrational modes strings can exhibit. This constraint is critical in giving rise to the diverse range of fundamental particles observed experimentally.

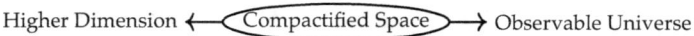

Figure 3.3: Schematic representation of a compactified dimension influencing observable particle physics.

Branes, on the other hand, are higher-dimensional analogs to strings. In theories such as M-theory, which extends from traditional string theory and proposes an 11-dimensional universe, branes can have up to 10 spatial dimensions. Just like strings, the dimensional makeup and dynamics of branes are crucial in shaping interactions at the quantum scale.

One must consider how branes and the dimensions they occupy interact. Branes are not merely passive entities; they are dynamic objects that can fluctuate, break apart, or join together. When strings end on branes or intersect with them, novel phenomena occur. These events lead to new physical implications, such as the generation of mass for gauge bosons (the particles responsible for force mediation such as photons and gluons), a process modeled effectively by connecting open strings to branes.

Moreover, the interaction between branes themselves can lead to significant cosmological phenomena, including the potential for brane collisions to explain cosmic inflation—an exponential expansion in the early universe. Each collision potentially reshapes the energy distribution in the cosmos, influencing both the macroscopic structure of the universe and its underlying quantum mechanics.

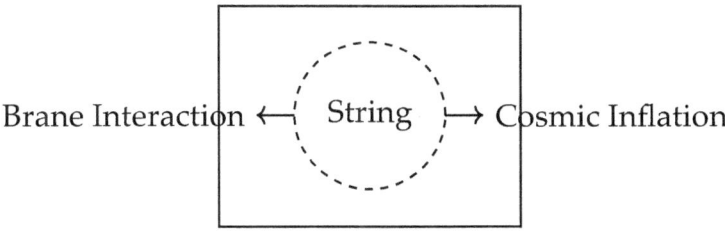

Figure 3.4: Interactions between strings, branes, and their implications for cosmic phenomena.

These intricate relationships between strings, branes, and the folded dimensions they inhabit provide not only a framework for understanding particle physics but also essential clues towards a unified theory of everything, harmonizing gravity with quantum mechanics and possibly leading to discoveries concerning dark matter and energy.

In exploring the full breadth of string theory and its implications on higher dimensional spaces, scientists pave a path towards potentially revolutionary insights into the very mechanisms that underpin reality itself, suggesting a universe far richer and more complex than our everyday experiences might suggest. This profound interconnectedness illuminated by string theory not only challenges our understanding but also invites us to rethink the very nature of space and time.

3.8 Vibrational Patterns of Strings and Their Implications

One of the most profound aspects of string theory is its postulate that the fundamental particles we observe are not point-like dots, but rather tiny, vibrating strings. These strings can vibrate in numerous modes, just as the strings on a violin or a piano have different vibrational patterns that produce different musical notes. Each mode of vibration of a string in string theory corresponds to a different elementary particle. The mass and charge of the particle are determined by the string's vibrational state. This correspondence suggests an elegant unification of all matter and forces in a single framework, predicated on the dynamics of these strings.

To delve deeper into these vibrational patterns, let us consider how a string can vibrate. A string's vibration can be mathematically described using Fourier series, which express the string's displacement as a sum of sinusoidal functions at various frequencies and amplitudes. In the simplistic model where we visualize a string as being one-dimensional and closed, like a loop, its vibrations can be depicted by:

3.8. VIBRATIONAL PATTERNS OF STRINGS AND THEIR IMPLICATIONS

$$y(x,t) = \sum_{n=1}^{\infty} A_n \sin(n\pi x/L) \cos(n\pi ct/L)$$
$$+ B_n \sin(n\pi x/L) \sin(n\pi ct/L)$$

where A_n and B_n are the amplitudes of the cosine and sine components of the vibration, L is the length of the string, c is the speed of light (reflecting the relativistic nature of the theory), x is the position along the string, and t is time. Each harmonic (specified by n) in this series corresponds to a different particle species.

The quantum behavior of these strings, which incorporates the principles of quantum mechanics into their vibrational patterns, is where the implications for particle physics become highly significant. In quantum string theory, each mode of vibration must obey the quantization condition which implies that only specific, discrete vibrational states are allowed. These states correspond to the quantum states of elementary particles.

Furthermore, considering the interaction between strings adds another layer to this complex picture. When two strings approach each other, they can either scatter off one another or merge into a single string and then later split apart again. These interactions mimic fundamental interactions between particles, such as electromagnetic and nuclear forces.

One intriguing implication of the vibrational pattern of strings is the unification of gravity with other fundamental forces. In string theory, gravity arises naturally from the vibrations of closed strings. As such, understanding these vibrations allows physicists to incorporate gravity into a unified framework alongside the other fundamental forces—a feat not achievable within the purview of traditional quantum field theories.

From a cosmological perspective, different vibrational patterns of strings during the early universe could have profound implications on the nature of cosmic inflation and on the types and concentrations of particles created during the Big Bang. Such scenarios could even influence the fabric of spacetime itself, leading to phenomena like black holes and the expansion of the universe.

Moreover, higher vibrational modes potentially reach into energy scales not yet observed experimentally. These modes could provide insights into physics beyond the Standard Model, including candidates for dark matter and energies associated with grand unified theories.

The analysis of vibrational modes is not just theoretical frivolity but forms the backbone of string theory's ability to model all known forces and particles in a singular theoretical framework. Researchers continue to explore these vibrational states through both mathematical innovations and experimental setups in particle physics, hoping to align these theoretical constructs with observable phenomena.

3.9 Interactions Between Strings and Branes

The interactions between strings and branes are pivotal in understanding the dynamic aspects of string theory. To delve deep into this subject, we must consider how strings, which are fundamentally one-dimensional objects, can interact with branes, which are entities of vari-

3.9. INTERACTIONS BETWEEN STRINGS AND BRANES

ous higher dimensions. As we analyze these interactions, it becomes clear that they are not just mere points of contact but complex processes that influence the stability, dynamics, and overall topology of the universe in string theory.

A fundamental interaction in this context is when a string ends on a brane, often referred to as an open string with endpoints attached to a brane. This scenario is crucial because the boundary conditions at the endpoints of the string must conform to the presence of the brane. Mathematically, these conditions are described by Dirichlet and Neumann boundary conditions, which respectively fix certain directions of the string's motion to lie within the brane and allow others to be free. The physical interpretation of these mathematical conditions brings to light the mechanisms by which strings can emit, absorb, or slide along branes.

To visualize this, consider a D-brane, which is a type of brane where open strings can have Dirichlet boundary conditions. When an open string ends on a D-brane, its endpoints are trapped within the D-brane. This setup allows the string to oscillate not only in the dimensions contained by the brane but also perpendicular to it, depending on the string mode's characteristics. The vibrational modes of these strings at the intersection with D-branes create various particle-like states which are integral in mediating forces and other physical phenomena within string theory.

When discussing higher dimensional interactions, such as those involving M-branes, the complexity increases. M-branes can support more exotic configurations like intersecting branes or branes within branes (commonly referred to as brane stacking). In these configurations, the interactions include not only the end points of strings but also scenarios where entire strings or membranes stretch between or wrap around multiple branes. These configurations are elegantly described using advanced mathematical tools like tensor calculus and differential geometry, which underscore the profound connection between physical phenomena and abstract mathematics in string theory.

Moreover, the exchange of virtual strings between branes illustrates another layer of interaction. Here, virtual strings, analogous to virtual particles in quantum field theory, can form temporary loops between branes, generating transient forces that could dictate brane motion or brane stability. This dynamic is akin to quantum fluctuations and is fundamental in hypothesizing about the non-static nature of the universe's fabric in string theoretical frameworks.

To support these theoretical models, graphical representations using multidimensional plots and topology diagrams can be extremely helpful. For instance, employing a tikz diagram to illustrate a string ending on a D-brane with appropriate boundary conditions can make the abstract concepts more tangible.

As we contemplate these interactions and their implications, we start to appreciate not only the theoretical beauty but also the potential practical implications these might hold for our understanding of quantum gravity and other unified force theories. The interactions between strings and branes tell a story of an intricately woven universe where dimensions beyond our perception shape the fundamental characteristics of everything we know about physics.

By weaving through these complex interactions and exploring their consequences, string theory continues to offer profound insights into the fabric of the cosmos, ensuring that our journey through these conceptual realms remains as intriguing as it is enlightening. The intricate dance between strings and branes in the higher-dimensional spaces of string theory provides a glimpse into the decidedly intricate yet beautifully symmetric structure of our universe.

3.10 Theoretical Predictions from Extra Dimensions

Delving deeper into the implications of string theory, a key aspect to consider is how the inherent existence of extra dimensions can lead to profound theoretical predictions. One critical characteristic of these predictions is their potential to solve several enduring puzzles in physics, such as the nature of dark matter, the fine tuning of cosmological constants, and the unification of forces.

Unification of Forces String theory operates under the premise that at the Planck scale—around 10^{-35} meters—all fundamental forces (gravitational, electromagnetic, weak, and strong) are unified. This unification is significantly influenced by the geometry and the topology of the extra dimensions. For instance, Calabi-Yau manifolds, a class of compact extra-dimensional spaces, often emerge in theories requiring supersymmetry. These manifolds shape the way strings can vibrate, and thus determine the types of particles and forces that are possible in the lower dimensional, observable universe.

Dark Matter In the framework of string theory, the extra dimensions may also offer natural candidates for dark matter. Particles known as axions, which are lightly interacting and substantially weak in mass, could theoretically leak into our observable universe from these hidden dimensions. The detection of such particles would not only offer a solution to the dark matter puzzle but also stand as a testament to the existence of extra dimensions.

Cosmological Implications Moreover, the theory of extra dimensions influences the cosmological constant problem—the discrepancy between the calculated and observed values of dark energy driving the universe's expansion. These dimensions could come into play by absorbing or redistributing the quantum fluctuations that contribute to cosmic expansion. Therefore, variations in the size and shape of extra dimensions over time may be an influential factor in models predicting the evolution of the universe's expansion rate.

Charting Particle Behaviour The geometric properties of extra dimensions can dramatically alter the paths and behaviors of subatomic particles. By integrating the effects of extra dimensions into particle physics models, researchers can craft novel predictions about particle interactions that could be observed at high-energy particle colliders, such as the Large Hadron Collider (LHC). For instance, certain extra-dimensional theories suggest deviations from the expected properties of the Higgs boson or predict new particles that could be detected in particle accelerator experiments.

Furthermore, plotting these predictions can be illustrated using Beacon rates in particle collision events within high-

3.10. THEORETICAL PREDICTIONS FROM EXTRA DIMENSIONS

energy physics experiments. The calculated deviations from Standard Model predictions can provide clear insight:

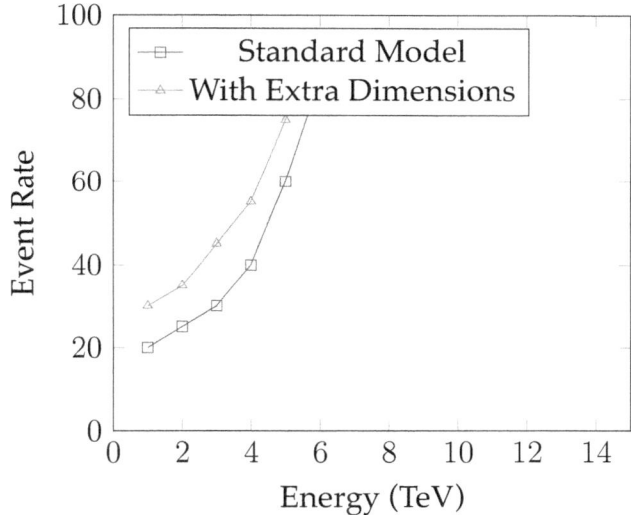

The theoretical implications of string theory's multi-dimensional framework enrich our understanding of both the cosmos and particle physics. The pursuit of experimental evidence for these predictions continues to drive forward not only the field of theoretical physics but also cutting-edge technological development in experimental setups.

Exploring this territory is akin to adding fine details to a magnificent tapestry. Each thread when viewed in isolation contributes a mere whisper to our comprehension. Yet, when woven intricately with others, reveals a breathtaking glimpse into the very fabric of our universe.

CHAPTER 3. FUNDAMENTAL CONCEPTS: STRINGS, BRANES, AND EXTRA DIMENSIONS

Chapter 4

Mathematical Underpinnings of String Theory

This chapter meticulously examines the intricate mathematics integral to string theory, encompassing a range of complex mathematical frameworks utilized to describe and predict phenomena within the theory. Topics such as Calabi-Yau spaces, conformal field theory, and supergravity are thoroughly explored, elucidating their critical roles in shaping the theoretical constructs of string theory. The chapter also addresses how mathematical concepts, like topology and algebraic geometry, contribute to our understanding of branes and the theory's attempt at describing the fundamental nature of reality.

4.1 Overview of Essential Mathematical Tools

In this section, we will delve deep into the arsenal of mathematical tools essential for understanding and progress in string theory. The primary focus will be on the foundational concepts and innovative methodologies that act as the backbone for more complex topics discussed in subsequent sections of this chapter.

Differential Geometry: Integral to any discussion on string theory is differential geometry. It involves the study of smooth curves, surfaces, and general multidimensional manifolds. In string theory, we explore the geometry of strings moving through multidimensional spaces, which are smoothly deformable and whose properties can vary from point to point. Understanding the curvature tensor, Riemannian metrics, and geodesics in differential geometry is vital for interpreting the dynamics of strings and the spacetime they inhabit.

For example, consider a manifold M potentially representing the spacetime fabric. Define its metric tensor g such that for any two tangent vectors u and v at any point in M, the metric $g(u, v)$ describes the geometric and physical properties, such as distances and angles, between these vectors.

Complex Analysis and Holomorphic Functions: These are crucial when dealing with conformal field theory aspects in string theory. A function is holomorphic if it is complex differentiable in a neighborhood of every point in its domain. This property leads to powerful results like Cauchy's integral formulas which are employed to calculate physical quantities in string systems.

For instance, the correlation functions in conformal field theory, a key topic in the analyses of string interactions, rely on the theory of complex variables to express these relationships. The modular invariance properties necessary for the consistency of string theories are testament to the far-reaching implications of complex analysis.

Algebraic Topology: The study of topological spaces with algebraic methods allows us to classify various structures that appear in theoretical physics. For string theory, concepts such as homology and cohomology groups are indispensable as they facilitate the understanding of branes and Calabi-Yau manifolds, which we will explore in later sections.

Consider the significance of homotopy groups, specifically fundamental groups, in analyzing possible mappings of a string's path in a looped space. Calculating these groups helps in understanding the allowable transformations and, hence, the basic physical characteristics of the space.

Quantum Mechanics and Path Integrals: An understanding of quantum behavior is essential given string theory's goal of unifying quantum mechanics and general relativity. The path integral formulation facilitates our grasp of the probability amplitudes for various states of a string. Each path represents a historical sequence of events that a string can undergo, and its amplitude contributes to the overall physical state as observed.

Lastly, we must integrate these mathematical tools with physical intuition to make sense of their outcomes. As we have adopted mathematical principles from various fields and applied them innovatively to challenges in theoretical physics, string theory continues to evolve. These mathematical methodologies do not merely support physical theories but often guide us to new predictions and insights that deepen our comprehension of the universe's fundamental fabric.

This detail provides a robust foundation for the subsequent exploration of specific applications within string theory, ensuring readers are well-equipped to tackle advanced concepts in later sections.

4.2 Calabi-Yau Spaces and Their Role in String Theory

Calabi-Yau spaces play a pivotal role in string theory, specifically in the compactification of extra dimensions proposed by the theory. To understand their importance, it is crucial to first grasp what these spaces are. Calabi-Yau manifolds are complex manifolds that have a vanishing first Chern class. This mathematical property implies that they can support Ricci-flat metrics, an essential requirement for consistency in superstring theory, which demands a 10-dimensional spacetime.

The compactification process, wherein dimensions are 'folded' into Calabi-Yau spaces, is not just about geometrically trimming down extra dimensions. It is about retaining the physical properties that allow superstring theory to be a viable candidate for unifying the four fundamental forces of nature. The idea is that while we perceive only four dimensions, additional dimensions are compacted so small — typically around the Planck scale — that they remain undetectable with current technology.

Let us consider the topology of these spaces more closely. A key property of Calabi-Yau manifolds is their ability to support non-trivial solutions to equations governing supersymmetry transformations. These are generally associated with special holonomy groups — particularly $SU(3)$ holonomy in the case of six real dimensions, which corresponds to three complex dimensions — crucial for the consistency of supersymmetric string theories.

Now, visualize entering a world smaller than particles, where dimensions curl up into complex geometric shapes. Here's where Calabi-Yau spaces become fascinating. They come in various shapes and sizes, each defined by a different complex structure and Kähler metric. The choice of these metrics defines the vibrational modes of strings, which in turn translate to the particle types and properties observed in the universe. A helpful analogy here might involve different musical instruments: Just as the shape and material of an instrument determine the notes it plays, so do the metrics of a Calabi-Yau space determine the 'notes' — elementary particles — that spring from vibrating strings.

Let's delve deeper into the impact of such configurations on physics. The rich structure of Calabi-Yau manifolds allows for a plethora of ways to 'wrap' branes around their cycles — where cycles refer to closed paths or looped paths that can be drawn on the manifold without lifting the pencil. Different wrappings correlate to different physical manifestations, such as charges and masses of elementary particles. These wrappings are not merely mathematical curiosities but are integral to predicting possible particle types and interactions possible within string theory frameworks.

For visual representation, imagine a multi-dimensional origami artwork. Just as altering the folds changes the final shape, slight alterations in the shape or size of a

Calabi-Yau space (or even the field configurations over it) leads to different physical properties at the macroscopic level, from particle masses to force charges.

To further reinforce this concept using mathematical visualization tools available in LaTeX, consider employing packages such as tikz for dynamic illustrations showing how varying these geometrical properties might influence observable physical phenomena.

As we peel the layers of string theory's mathematical complexities back, it becomes increasingly clear: Calabi-Yau manifolds are not merely esoteric mathematical constructs but are central to bridging gaps between budding theoretical predictions and tangible, observable aspects of our universe. Through them, string theory endeavors to sketch a finer picture of the universe's fundamental fabric, helping to stitch together the cosmos's seemingly disparate forces into a unified theory.

4.3 Topology and Manifolds: Setting the Stage for Extra Dimensions

Discussing the profound impact of topology and manifold theory within string theory necessitates a deep dive into their roles in conceptualizing the theoretical framework that supports extra dimensions. Topology, essentially the study of properties preserved through deformations, stretching, and twisting of objects, bridges an important link with manifolds in string theory - providing a mathematical lens through which the elegance of extra dimensions can be examined without the limitations imposed by conventional three-dimensional space.

4.3. TOPOLOGY AND MANIFOLDS: SETTING THE STAGE FOR EXTRA DIMENSIONS

A manifold, in the context of string theory, is a mathematical space that, on small scales, resembles the Euclidean space of a specific dimension. As entities that can be curved and possess a more complicated global structure, they are central to the theory's narrative of extra dimensions. These additional dimensions are not observable at our everyday scale, but they are theorized to exist at minuscule scales, curled up in the fabric of the universe.

Differentiating Manifolds: To fully appreciate their role in string theory, one needs to understand two key types of manifolds: compact and non-compact. Compact manifolds are finite and without boundary, akin to the surface of a sphere. Non-compact manifolds, like Euclidean space, extend infinitely. String theory often employs compactified dimensions that are modeled using compact manifolds due to their finite nature which succinctly explains why these extra dimensions are not observed in our daily experiences.

Calabi-Yau Spaces: Among the most significant types of manifolds in string theory are Calabi-Yau spaces. These spaces facilitate the compactification process necessary for superstring theory to consolidate its ten required dimensions into the four observable ones. The properties of Calabi-Yau manifolds, especially their ability to support structures like Ricci-flat metrics, makes them indispensable in the realm of string theory, echoing the subtle nuance of mathematical and physical interplay.

Mathematical Representation: We leverage a variety of mathematical tools to describe these manifolds. Consider the representation given by a simple manifold equation in local coordinates:

$$M = \{(x, y, z) \in \mathbb{R}^3 : x^2 + y^2 + z^2 = 1\},$$

which describes a three-dimensional sphere, a compact

manifold.

Further, exploring differential geometry, the study of curves, surfaces, and general manifolds provides techniques to describe spaces where string theory's extra dimensions 'live'. Here, we use tensors to represent physical and geometric quantities such as curvature, which is intimately related to the gravitation in string theory.

Curvature and String Theory: The curvature of manifolds within string theory illustrates how these shapes influence string dynamics. The Riemann curvature tensor, a key tool in this framework, encapsulates how much a manifold deviates from being flat. Its implications reach far into the gravitational effects predicted by string theory, providing a core component of the theoretical structure.

Allowing manifolds and their mathematical portrayal to guide our understanding, we grasp a more cohesive picture of the theoretical backdrop string theory proposes. This mathematical structure is not merely a background for physical phenomena; it actively shapes the predictions and fundamentals of the theory.

Moving ahead, the implication of this intricate relationship between mathematics and physics serves not only as a theoretical development but also as a beacon guiding practical computational techniques and simulations in string theory research. Through this exploration of topological spaces and manifold structures, we uncover further the enigma of dimensions beyond our direct perception - all encoded subtly in the mathematical fabric woven by string theory.

This section extensively discusses how topology and manifolds contribute to the setup of extra dimensions in string theory. From different types of manifolds and their characteristics to their implications in understanding curvature and gravity within string theory's framework is detailed, providing deep insights into how these mathematical concepts enrich and propel theoretical physics in the realm of higher dimensions.

4.4 Conformal Field Theory in String Dynamics

Conformal Field Theory (CFT) is a cornerstone in the structure of string theory, providing powerful algebraic techniques that underpin much of the theory's description of the fundamental forces and particles in the universe. This mathematical framework uniquely suits the needs of string theory due to its ability to maintain the shape of the equations of motion under conformal transformations, which essentially preserve angles but not necessarily distances.

The motivation for using CFT in string theory can mainly be traced back to the need for a consistent and anomaly-free theory that operates in two dimensions— the world-sheet dimensionality of strings in spacetime. The critical feature here is the string world-sheet conformal invariance, which implies that physical predictions must remain invariant under local scaling transformations of the world-sheet metric.

To properly appreciate the role of CFT in string dynamics, consider a flat two-dimensional surface representing a string's world-sheet. The metrics on this world-sheet can experience local transformations that alter distances but retain angles; the transformations are what mathematicians refer to as conformal. These transformations form a group known as the conformal group. In two dimensions, this group is infinitely dimensional, which significantly enriches the structure and solubility of associated field theories.

$$ds^2 \to \lambda(z, \bar{z})ds^2$$
$$z \to f(z), \ \bar{z} \to \bar{f}(\bar{z})$$

Here, ds^2 represents an infinitesimal distance on the world-sheet, $\lambda(z, \bar{z})$ is a scaling function, and z, $\bar{z}$ are complex coordinates on the world-sheet. The functions $f(z)$ and $\bar{f}(\bar{z})$ demonstrate how complex transformations manifest under the conformal group.

The conformal anomaly (or central charge) plays a crucial role, especially in determining the consistency of string theories through different spacetime dimensions. For example, the critical dimension where string theories are most naturally anomaly-free is 26 for bosonic strings and 10 for superstrings. These are largely the results derived using CFT tools:

$$c - 26 = 0 \quad \text{(Bosonic)}$$
$$c - 10 = 0 \quad \text{(Superstring)}$$

Where c is the central charge associated with the conformal algebra on the world-sheet. One notable outcome from the algebra of terms for various fields on this world-sheet is the Virasoro algebra, an extension of the conformal algebra that includes an infinite sequence of nontrivial central extensions— essentially introducing 'memory' of transformations' past configurations.

4.4. CONFORMAL FIELD THEORY IN STRING DYNAMICS

$$[L_m, L_n] = (m-n)L_{m+n} + \frac{c}{12}(m^3 - m)\delta_{m+n,0}$$

The fields in CFT on the world-sheet are primary fields, and descendant fields expressed in terms of 'operator product expansion' (OPE). These expansions are crucial in navigating interactions in CFT since they provide a systematic way of understanding the effects of field interactions localized at points on the world-sheet:

$$\phi_i(z)\phi_j(w) \sim \sum_k \frac{C_{ij}^k \phi_k(w)}{(z-w)^{h_i+h_j-h_k}}$$

Lastly, the subtlety and complexity of CFT find an elegant application in characterizing various topological and dynamical aspects of string theory, including the classification of possible string backgrounds based on allowed conformal field theories. These backgrounds serve as potential universes in string theory cosmology and motivate diverse solutions to string dynamics problems, thereby enriching our understanding beyond conventional physics interpretations.

Thus, through CFT, we delve into a deep mathematical portrayal that not only enriches our comprehension of string dynamics but also connects closely with various symmetries essential in higher-dimensional theories, paving the way for insightful discoveries in theoretical physics.

4.5 Supergravity in Higher Dimensions

Supergravity, a theoretical framework aiming to unify general relativity with supersymmetry, plays a pivotal role in string theory, particularly when extended into higher dimensional spaces. Unlike traditional four-dimensional supergravity, higher dimensional versions introduce unique and complex structures that are instrumental in understanding the unification of all fundamental forces and the underlying structure of spacetime posited by string theory. This exploration requires a robust engagement with differential geometry and group theory, as these mathematical arenas provide the necessary tools to describe how supergravity operates in dimensions beyond our conventional experience.

The necessity of considering higher dimensions arises from the inherent requirements of string theory. String theory posits that point-like particles are replaced by one-dimensional strings whose modes of vibration correspond to various particle types. The theory elegantly resolves several theoretical conundrums, including the integration of gravity with quantum mechanics, by proposing additional spatial dimensions. These extra dimensions are typically compactified on Calabi-Yau spaces, which leads to the consideration of supergravity in these higher-dimensional compactifications.

In supergravity theories in higher dimensions, such as 11-dimensional supergravity which is the maximal extension possible under M-theory, the spacetime is described by a metric that encodes gravitational interactions. The metric's compatibility with supersymmetry introduces further complexity, requiring the introduction of addi-

4.5. SUPERGRAVITY IN HIGHER DIMENSIONS

tional fields and constraints. These include the gravitino, a supersymmetric partner of the graviton, carrying spin 3/2, which couples to the metric and other fields in a way that respects both supersymmetry and Lorentz invariance.

To analyze the dynamics and implications of these high-dimensional theories, we turn to techniques such as dimensional reduction and compactification. Dimensional reduction is a method where extra dimensions are mathematically reformed to manifest as additional fields or altered physical constants in lower dimensions. Compactification, on the other hand, entails folding the extra dimensions into compact manifolds small enough to be unobservable at current energy scales, with Calabi-Yau manifolds being a prime example due to their rich geometric structure and ability to preserve supersymmetry.

Using advanced mathematical tools, such as tensor analysis and group theory, one can discern the interactions among the fields embedded in the fabric of higher dimensions. Specifically, the rich interplay between Riemannian geometry and Yang-Mills theories elucidates how forces other than gravity—electromagnetic, strong and weak—are unified in a higher-dimensional setting. The curvature of these manifolds, their topological features, and their group actions are fundamental in revealing how these forces might emerge from a singular unified field theory.

Moreover, exploring anomaly cancellation within these theories highlights their mathematical consistency and physical viability. Anomalies — inconsistencies that can arise when trying to quantize a classical theory — pose significant challenges. In supergravity, one must carefully examine the potential anomalies that can occur due to the presence of higher spin fields and their interactions with the manifold's topology. The Green-Schwarz mechanism is one particular method employed to tackle anomaly issues, ensuring that supergravity remains a consistent and viable theory within higher-dimensional frameworks.

As we explore these detailed constructions, it becomes apparent that supergravity not only enriches our understanding of fundamental physics but also provides a compelling narrative for the fabric of the cosmos. It underscores the sophistication and elegance with which the universe might be woven, a seamless blend of mathematical rigor and physical insight that transcends our conventional understanding of space and time.

This exploration into supergravity not only nourishes our technical arsenal but also broadens our conceptual horizon, preparing us for further inquiry into the profound implications of string theory in describing the universe at its most fundamental level.

4.6 Algebraic Geometry: Essential Techniques for Brane Construction

Algebraic geometry, a profound cornerstone in the mathematics of string theory, supplies indispensable tools for the formalization and understanding of brane constructions. Branes—fundamental objects within string theory—require an advanced mathematical description that algebraic geometry particularly provides through its exploration of zeros of multivariate polynomials and

4.6. ALGEBRAIC GEOMETRY: ESSENTIAL TECHNIQUES FOR BRANE CONSTRUCTION

their resulting structures.

The quintessence of algebraic geometry lies in its ability to handle complex spaces, or varieties, which are solutions to systems of polynomial equations. Within the context of string theory, these varieties often manifest as the extra dimensions of space suggested by the theory. Branes themselves can be modeled as subvarieties within these higher-dimensional spaces, making the role of algebraic geometry not just relevant but essential.

Consider the typical form of a Calabi-Yau manifold, a space that locally might be represented as a solution to a polynomial equation like $X^n + Y^n + Z^n = 1$. The role of algebraic geometry in describing such spaces involves sophisticated tools such as sheaf theory, schemes, and cohomology. These tools allow for the detailed study of the manifold's properties like symmetry and singularity, which are crucial for the stability and dynamics of branes in string theory.

In the operational framework of string theory, D-branes, which are types of branes supported by open strings, are particularly influenced by the algebraic topology of the embedding space. The precise construction of D-branes in algebraic terms involves the characterization of their worldvolume - essentially the 'surface' or 'space' that the brane occupies - using coherent sheaves. A coherent sheaf over a variety like a Calabi-Yau manifold can effectively model how fields associated with the brane vary over the manifold.

To elaborate on this with an example, consider a D-brane wrapping around a complex submanifold M of a Calabi-Yau space X. The distribution of fields on the D-brane can be represented by a coherent sheaf $\mathcal{F}$ on X, localized on M. The sheaf $\mathcal{F}$ gives a rigorous way of discussing bundles and their sections without resorting to local triviality, thus accommodating singularities and other complex geometrical features naturally occurring in string theory scenarios.

Moreover, the Chern classes, fundamental constructs in algebraic geometry, provide critical information about the embedding of D-branes in these higher-dimensional manifolds. These classes can determine the topological characteristics of the brane configurations, which are essential for understanding their physical implications, such as charge, tension, and interaction properties.

Another illustrative aspect is the use of derived categories in algebraic geometry. Derived categories facilitate dealing with complex objects like branes as they allow for the construction of various homological invariants that can describe more convoluted interactions than is possible with traditional cohomological techniques. For instance, the derived category of coherent sheaves on a Calabi-Yau manifold helps in understanding how branes can be transformed or deformed within the manifold, a concept crucial for exploring dualities in string theory.

In providing a clear framework for these operations, algebraic geometry aids not only in modeling the physical aspects of branes but also serves as a bridge to quantum field theories through the elegant treatment of gauge symmetries and other field-theoretic constructs. This compatibility between string theory and other fields of physics underscores the extradimensional versatility of algebraic geometry.

The motion within this mathematical landscape offers an increasingly precise blueprint for understanding our universe's very fabric, demonstrating how string theory and its reliance on algebraic geometry expand not only the bounds of theoretical physics but also our capability to conceive realms beyond tangible perception.

4.7 Quantum Geometry and Its Implications for String Theory

Quantum geometry forms a cornerstone in unraveling the fabric of string theory, providing profound insights and tools necessary for understanding the structure of space-time at subatomic scales. An imperative aspect of quantum geometry within string theory is its role in the formulation of string theory's spacetime, which is intrinsically quantum in nature due to the minuscule scale at which strings operate.

At the heart of quantum geometry lies the concept of non-commutative geometry, which posits that the coordinates defining the positions of points in space do not commute. In mathematical terms, if x and y are coordinates on a plane, then in classical geometry, $xy = yx$. However, in a non-commutative space, $xy \neq yx$, introducing an uncertainty akin to the Heisenberg uncertainty principle in quantum mechanics. This paradigm shift, from commutative to non-commutative coordinates, leads to a host of novel mathematical structures and theories, which are instrumental in describing the quantum behaviors of strings.

To explore this further, consider how this non-commutativity manifests within the algebraic structure known as the non-commutative torus. This model serves as an ideal playground for theoretical exploration, provided by its simplicity and its rich structure. The non-commutative torus is described by the algebra $\mathcal{A}_\theta$, where θ is a parameter representing the non-commutativity. Elements of $\mathcal{A}_\theta$ can be visualized as functions on a regular torus, but with the multiplication twisted by θ.

Mathematically, this is represented by:

$$f * g(x) = \sum_{k \in \mathbb{Z}^n} e^{2\pi i \theta k \cdot m} \widehat{f}(k)\widehat{g}(m-k),$$

where f, g are elements (functions) in $\mathcal{A}_\theta$, and $\widehat{f}, \widehat{g}$ denote their Fourier transforms. This product $*$ introduces a deformation in the conventional function multiplication, leading to modified physics at quantum levels.

By utilizing this non-commutative framework, string theorists have postulated models whereby the spacetime fabric itself is fundamentally quantum mechanical in nature, yielding predictions and insights into the behavior of strings under various physical and cosmological conditions. For instance, when examined under the lens of quantum geometry, the interaction among multiple strings or the self-interaction of a single string may be represented as topological transformations within these non-commutative spaces.

Furthermore, quantum geometry has spurred significant advancements in understanding dualities in string theory. Intricately connected to mirror symmetry—that connects two distinct Calabi-Yau spaces—quantum geometric approaches provide a robust mathematical framework to analyze and predict how physical properties transpose between dual theories. Specifically, the topological vertex, a concept born from quantum geometry, allows computation of string amplitudes via its representation as vertices in a Feynman-like diagrammatic approach.

Gazing ahead into the vista of uncharted territories within string theory, it is evident that quantum geometry will continue to shape not only our mathematical frameworks but also influence the conceptual methodologies we employ to comprehend the universe's most fundamental aspects. By elegantly fusing geometrical concepts with quantum mechanics, it offers an insightful paradigm through which the cosmic tapestry, woven by strings, can be further explored and understood.

4.8 Supersymmetry: Bridging Mathematics with Physical Phenomena

Supersymmetry stands as a pivotal mathematical concept in string theory, offering fascinating insights into the symmetrical relationships between bosons and fermions—two fundamental particle classes in the universe. This symmetry suggests that each boson, a particle responsible for carrying forces, has a corresponding partner fermion, which makes up matter. By postulating these symmetrical pairings, supersymmetry not only aims to unify the forces and matter in a single theoretic framework but also resolves several pressing issues in modern physics, such as the hierarchy problem which concerns the vast difference in strength between the gravitational force and other fundamental forces.

One of the key mathematical tools employed in supersymmetry is the use of superalgebras. These extend the concept of a Lie algebra (which underpins much of the mathematical structure of particle physics) by incorporating generators that satisfy anticommutation relations in addition to the usual commutation relations. In the formalism of supersymmetry, these new generators facilitate transformations between fermionic and bosonic states, providing a deep algebraic structure that supports the symmetrical nature of all fundamental particles and interactions.

The representation theory of these superalgebras, particularly important in supergravity—a supersymmetric extension of general relativity—compels a rethinking of spacetime geometry. In supergravity, the addition of supersymmetric partners to the graviton (the hypothetical quantum field carrier of gravity) into the model introduces higher dimensions of space wherein these particles can exist without violating established laws of physics.

One can visualize the concept of supersymmetry in string theory through the exploration of higher-dimensional spaces.

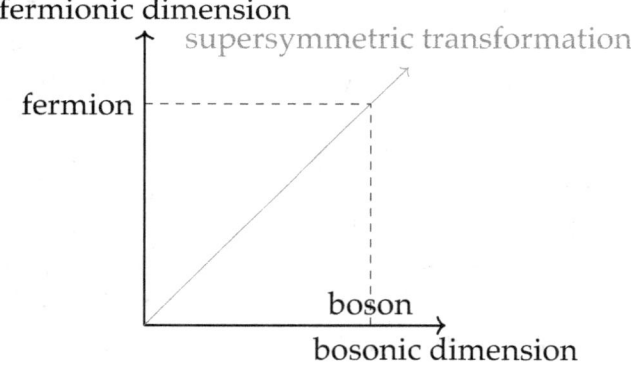

Applications of supersymmetry extend beyond theoretical constructs; they offer promising methods for understanding particle physics phenomena observed in experiments such as the Large Hadron Collider. It is hypothesized that superpartners to known particles could be produced in high-energy collisions if supersymmetry is indeed a reality of our universe. Analyzing the potential

4.8. SUPERSYMMETRY: BRIDGING MATHEMATICS WITH PHYSICAL PHENOMENA

detection of these superpartners helps guide experimental designs and interpret their results, establishing a direct line of inquiry from abstract mathematics to observable physics.

Furthermore, by requiring supersymmetry, string theory maintains greater internal consistency when used to explore extraordinarily high-energy realms where traditional gravity ceases to work effectively alone. This necessitates the inclusion of supersymmetric particles and forces into any comprehensive model designed to unify all known physical interactions under one overarching theoretical framework.

While immensely powerful and aesthetically appealing due to its symmetrical properties, supersymmetry also faces challenges such as lack of empirical evidence for superpartners. This continues to be a vital area of research, with physicists and mathematicians both striving to refine theoretical predictions and align them with experimental data.

As new experimental technologies and theoretical advancements emerge, the continued exploration of supersymmetry remains essential in our quest to uncover the deeper symmetrical nature of the universe. Through the lens of supersymetry, both mathematicians and physicists may discover the threads necessary to weave together the disjointed patches of our current understanding into a complete and harmonious depiction of reality.

4.9 Dualities and Symmetries: Mathematical Magical Mirrors

In the exploration of string theory, dualities and symmetries perform much like magicians, revealing unexpected connections and transformations between seemingly disparate areas of physics and higher-dimensional mathematics. One of the crowning features of dualities is their ability to provide profound insights into the equating of theories with different physical parameters, suggesting equivalences that are both startling and illuminating.

What are Dualities? At its core, a duality in string theory implies that two different theories can be descriptions of the same phenomena under different physical conditions. This is crucial because it allows physicists to use the simpler of two dual theories to answer questions within a more complex framework. The primary kinds of dualities explored in string theory include T-duality and S-duality.

- **T-duality** heightens our understanding by suggesting that a string wrapping around a compact dimension of radius R can be equivalently described by a string wrapping around a radius $\frac{\alpha'}{R}$ where α' is the string tension. This insight suggests a fascinating symmetry between large and small-scale structures in string theory.

- **S-duality**, on the other hand, intertwines the strength of coupling constants. It relates a theory with a strong coupling constant to a theory with a weak coupling constant, thereby bridging gaps in computational feasibilities and offering a broader spectrum of theoretical exploration.

The Role of Symmetry: The concept of symmetry in string theory extends beyond mere aesthetic appeal; it forms the bedrock of conservation laws and invariant properties under transformations — components that are essential for the consistency and solvability of the theory. Symmetry in string theory generally manifests in several ways, including Lorentz invariance within the string, and gauge symmetries arising from open string endpoints.

Mirror Symmetry: Among the jargon of string theory, mirror symmetry stands particularly noteworthy. It operates primarily within the realm of Calabi-Yau spaces, positing that pairs of these compactifications can result in identical physics despite their geometric dissemblance. This kind of symmetry underlines the deep geometrical and topological root of string theory, mapped effectively using algebraic geometry tools.

Mathematical Implication and Visualization: With dualities painting a rich tapestry of interconnected theories, visualizing these concepts mathematically elevates understanding. For example, consider a plot visualized using TikZ or pgfplots in LaTeX to illustrate the relationship between the radius of compactification R and its reciprocal in T-duality.

CHAPTER 4. MATHEMATICAL UNDERPINNINGS OF STRING THEORY

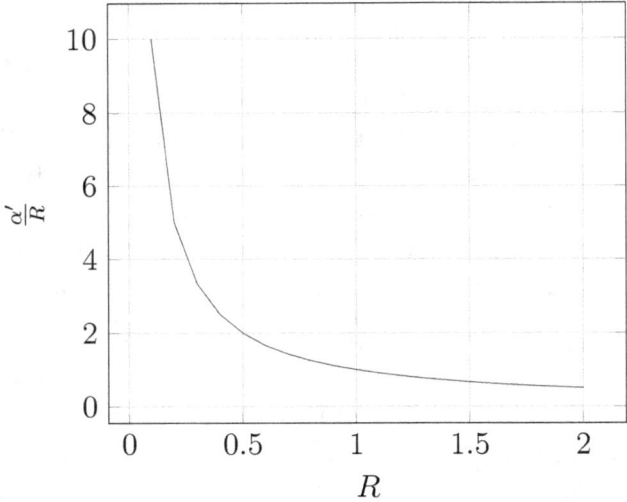

Dualities escape rigid frameworks, suggesting a cosmos where distances, strengths of interaction, and even dimensions are not static or absolute but are relative and interconnected through mathematical symmetries. The ability to switch between descriptions highlights a larger theoretical flexibility and coherence, hinting at the potentially universal nature of string theory across differing scales and forces in the universe.

Exploring further into dualities and symmetries helps not only in evolving string theory itself but also enriches the associated mathematical disciplines, bringing together disparate mathematical techniques under unified frameworks that have as much elegance as utility. This intersection inspires advancements in numerous fields, blending abstract mathematical concepts with tangible physical theories — a testament to the underlying unity of all things within the fabric of cosmos.

4.10 Applying String Theory Mathematics to Other Fields of Physics

The mathematical frameworks developed for string theory are not confined solely to the realm of theoretical physics. Indeed, the powerful tools and concepts have found numerous applications in disparate areas of physics, enabling new insights and methodologies that enrich our understanding of the physical world.

A notable extension of string theory mathematics is seen in the domain of quantum computing and information. The intricate relationship between entanglement and geometry in string theory has prompted a fruitful exchange of ideas with quantum information theory. For instance, the concept of holographic entanglement entropy, derived from the holographic principle in string theory, has analogs in the computation of entanglement measures in quantum systems. This cross-pollination has led to a deeper understanding of quantum entanglement properties and has implications for designing algorithms and error-correction schemes in quantum computers.

Further, string theory's mathematical toolset has also enriched the field of condensed matter physics. Concepts such as AdS/CFT correspondence — a duality between a type of string theory and a quantum field theory — provide a novel framework for studying strange metals and high-temperature superconductors. These materials, critical in the development of advanced technological applications, exhibit properties that are challenging to study using traditional methods. The correspondence offers a powerful theoretical lens to examine quantum criticality and phase transitions, providing predictions and insights that have been corroborated experimentally.

In cosmology, the mathematical rigor of string theory has given rise to new models and theories concerning the early universe, particularly in the context of inflation theory. String theoretic concepts such as brane inflation — where our universe is envisioned as a brane within a higher-dimensional space — offer unique explanations for the rapid expansion of the early universe. This approach not only provides a fertile ground for theoretical exploration but also ties closely with observational data, guiding the development of experiments and the interpretation of cosmic microwave background radiation measurements.

The influence extends into the chaotic systems and nonlinear dynamics, where techniques from string theory are utilized to address problems in turbulence and complex systems. The mathematics of string theory, particularly related to Riemann surfaces and conformal mappings, provides tools for understanding the behavior of systems far from equilibrium, which is a longstanding challenge in theoretical physics.

To illustrate these applications, consider the potential energy landscape in quantum field theories, which can be extraordinarily complex. String theory introduces sophisticated geometric and topological methods that simplify the visualization and calculation of these landscapes, offering new avenues to tackle problems in particle physics beyond the Standard Model.

The impact of string theory's mathematical structures in physics is profound and far-reaching. It exemplifies a successful interplay between pure mathematical curiosity and practical physical inquiry. As research continues, it is anticipated that the influence of string theory mathematics will pervade even more areas of physics, potentially

4.10. APPLYING STRING THEORY MATHEMATICS TO OTHER FIELDS OF PHYSICS

leading to groundbreaking discoveries and innovations.

This seamless embedding of string theory mathematics into various fields of physics not only highlights its versatility and power but also underscores the unity of scientific inquiry — an endeavor that continually seeks to describe the diverse aspects of our universe through a common language of mathematics.

CHAPTER 4. MATHEMATICAL UNDERPINNINGS OF STRING THEORY

Chapter 5

Quantum Mechanics and String Theory: A Symbiotic Relationship

This chapter explores the deep interconnections between string theory and quantum mechanics, emphasizing how fundamental concepts from quantum mechanics are woven into the fabric of string theory. It covers how string theory extends and integrates these ideas to explain the behavior of strings and their interactions, addressing phenomena such as quantum entanglement, fluctuations, and coherence within the framework of string theory. Additionally, the chapter discusses the implications of this synthesis for understanding quantum gravity and highlights the theoretical advancements derived from blending these two pivotal fields of physics.

5.1 Basics of Quantum Mechanics: A Quick Overview

Quantum mechanics stands as a cornerstone of modern physics, elucidating the behavior of matter and energy at the smallest scales. Its introduction brought a profound shift from classical mechanics, delivering equations that allow us to predict how particles like electrons and photons behave.

The key equation underpinning quantum mechanics is the Schrödinger equation, described as:

$$i\hbar \frac{\partial \psi}{\partial t} = H\psi$$

where ψ represents the wavefunction of the system, H is the Hamiltonian operator (describing the total energy of the system), $\hbar$ is the reduced Planck constant, and t denotes time. This equation is vital for determining the probability amplitudes of a particle's position and other physical properties.

Central to quantum mechanics is the concept of the wavefunction, which provides a probability distribution for the position, momentum, and other physical properties of a particle. Unlike classical predictions, quantum mechanics describes properties in terms of probabilities, leading to the fundamental nature of uncertainty in the subatomic world. This is formally articulated through Heisenberg's uncertainty principle:

$$\Delta x \Delta p \geq \frac{\hbar}{2}$$

where Δx and Δp denote the uncertainties in position and momentum, respectively. This principle implies

5.1. BASICS OF QUANTUM MECHANICS: A QUICK OVERVIEW

that more precise measurement of one property leads to greater uncertainty in the other.

Quantum mechanics also introduces the concept of superposition, where a particle exists simultaneously in multiple states until it is measured. This phenomenon is famously encapsulated in Schrödinger's cat paradox, where a cat in a sealed box is described as being simultaneously alive and dead until the box is opened and the state "observed."

Furthermore, quantum entanglement presents another peculiar aspect of quantum mechanics, where particles become so deeply connected that the state of one (no matter how far apart) instantaneously influences the state of another. This has profound implications for quantum information theories, including quantum computing and cryptography.

To visualize some applications and phenomena described by quantum mechanics, consider using diagrams or simple graphs. For instance, plotting a wavefunction $\psi(x)$ could illustrate changes in probability densities across space. Here, a simple Gaussian wave packet could be visualized:

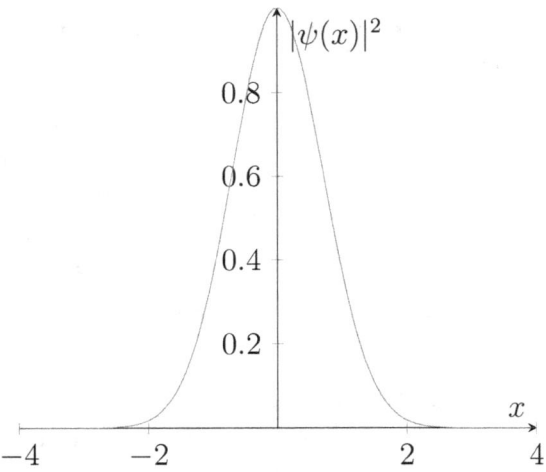

In sum, quantum mechanics provides a framework that remarkably explains and predicts the behavior of systems at microscopic scales. As we transition to discussing how these principles integrate with string theory, we continue to build on the solid foundation established by these quantum characteristics, aiming to solve more complex phenomena in physics.

5.2 Integrating Quantum Mechanics with String Theory

Delving into the integration of quantum mechanics within string theory entails a meticulous exploration of how quantum principles sculpt the understanding of string dynamics and their multiple degrees of freedom. The correspondence starts at a fundamental level, where the probabilistic nature of quantum mechanics introduces uncertainty into the precise characterization of strings, a core aspect that shapes the overall behavior of these

5.2. INTEGRATING QUANTUM MECHANICS WITH STRING THEORY

fundamental entities in the universe.

In classical string theory, a string's position and state are determined by mathematical expressions that describe its shape and vibrations. However, like particles in quantum mechanics, strings exhibit dual characteristics, displaying both particle-like and wave-like behaviors. This duality is crucially incorporated into string theory by modeling strings not merely as fixed one-dimensional objects in space-time but as probabilistic entities described by wave functions.

The wave function in quantum mechanics, denoted often by $\psi(x,t)$ for a particle, specifies the probability amplitude for finding a particle at a position x at a time t. Analogously, string theory extends this concept to strings, wherein the wave function Ψ now depends on the entire configuration of the string itself, encapsulating a vastly more complex parameter set that includes all possible modes and shapes of the string. The Schrödinger equation in quantum mechanics, governing the evolution of wave functions, finds its counterpart in the functional Schrödinger equation for strings. This equation governs how the wave function of a string configuration evolves with time.

$$i\hbar \frac{\partial}{\partial t}\Psi = H\Psi$$

where H represents the Hamiltonian operator for strings. Directly modifying and applying quantum mechanical operators within this context, necessitates sophisticated mathematical tools used in quantum field theory and adapted uniquely to the needs of string dynamics.

Given that each point on a string can interact in ways particles can, but now extended over its length, interactions become exceedingly complex and are governed by what is known as string field theory. Within this framework, individual strings can split and join, encapsulating the quantum mechanical phenomena of entanglement and superposition on a magnified scale. Such interactions echo the quantum principles where the outcome of measurements becomes intrinsically probabilistic, demanding an ensemble of possible string configurations to predict phenomena accurately.

Moreover, integrating quantum mechanics has profound implications on understanding string dynamics at different energy scales. At high energies, close to the Planck scale, the quantum nature of strings becomes paramount as Heisenberg's uncertainty principle dictates that precise measurements of positions become increasingly fraught with greater uncertainties. This is particularly relevant in scenarios involving small-scale, high-energy interactions where traditional concepts like locality begin to blur, and more holistic, non-local descriptions provided by string theory take precedence.

To graphically visualize how quantum mechanics influences string behavior, consider the vibrations of a string. Each modal vibration, in essence, could be treated akin to the quantum harmonic oscillator. Here, each mode's energy levels can be quantized, leading to discrete energy states. Employing a plot to demonstrate the energy states versus vibrational modes would illustrate the quantum mechanical nature expressed in tangible string dynamics.

The synergy between quantum mechanics and string theory not only enriches understanding but also magnifies our perspective on the universe's fundamental workings, providing a more unified framework that might one day lead to uncovering all-encompassing laws governing everything from subatomic particles to cosmological phenomena. This powerful fusion continues to inspire new theoretical ideas and experimental approaches aiming at discovering underlying principles that remain concealed within conventional frameworks.

5.3 Quantum Field Theory and Its Connection to Strings

Quantum Field Theory (QFT) serves as a cornerstone in formulating the principles of particle physics and has profound implications in the realm of string theory. To understand how QFT integrates with string theory, we first revisit the concept that in QFT, particles are excitations of underlying fields. For instance, the electromagnetic field, when quantized, manifests photons, and similarly, other fields correspond to particular particles.

The leap from traditional particle-based QFT to string theory involves extending the notion of zero-dimensional point particles to one-dimensional objects known as strings. These strings can oscillate, stretch, join or split, encompassing a richer variety of behaviors than point particles. The mathematical transition from points to strings introduces significant complexity but also a richer framework for describing fundamental interactions.

Let us take a closer look at the perturbative aspects of string theory and how they relate to QFT. The perturbative treatments in both theories involve calculating the effects of small perturbations around a fixed background. In QFT, the calculations often revolve around Feynman diagrams, which represent interactions between particles as vertices connected by lines (propagators). In string theory, these diagrams evolve into two-dimensional surfaces known as world-sheets, whose boundaries trace out the paths of strings in spacetime.

A crucial area where QFT influences string theory is in the renormalization process. Renormalization, necessary in QFT due to infinities arising in calculations at small scales, also emerges in string theory but in a different guise. Strings have a natural cut-off scale due to their tension and finite size, which softens the ultraviolet divergences typical in point-like particle theories. This inherent property of strings grants them an advantage, potentially offering a route to a finite theory of quantum gravity.

Another pivotal concept is gauge symmetry, originally rooted in QFT. In string theory, gauge symmetries arise naturally from the vibrational modes of the string. For instance, the Type I string theory contains both open and closed strings, where the open strings' ends adhere to D-branes, carrying gauge interactions akin to those described by QFT, thus reflecting a profound connection between both theoretical frameworks.

In the dynamic domain of string interactions, QFT plays another essential role through the technique of effective field theories (EFTs). EFTs allow physicists to focus on phenomena at accessible energy scales, disregarding the higher energy processes. Similarly, in string theory, one can derive an effective field theory by considering only low-energy string modes. This gives rise to predictions that can be compared with experimental data within the accessible energetic realms, bridging a gap between highly abstract string phenomena and observable physics.

To encapsulate these connections visually, consider a diagrammatic comparison between Feynman diagrams in QFT and world-sheet descriptions in string theory. This can be achieved using a simple transition from point interactions expanding into linear world-sheet interactions encapsulated by:

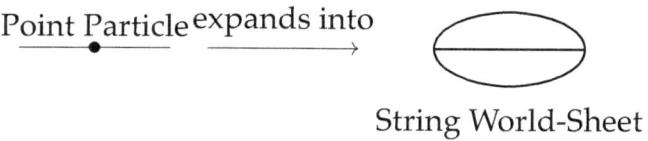

String World-Sheet

This diagram underscores the expansion of conceptual frameworks from QFT to string theory where interactions are not merely points but complex geometrical constructions in higher dimensions.

By unraveling these intricate relationships between QFT and string theory, we gain a deeper understating not only of the strings themselves but also of the fundamental forces they embody. It paints a promising picture of a unifying theory potentially capable of addressing some of the most perplexing questions in modern physics.

5.4 The Role of Quantum Mechanics in Describing String Interactions

Understanding the role of quantum mechanics in describing string interactions calls for a nuanced grasp of both quantum principles and the core tenets of string theory. String theory suggests that fundamental particles are not point-like dots, but rather elongated strings whose vibrational patterns determine the observed properties such as mass and charge. These strings can oscillate, split, and combine, leading to a rich variety of physical phenomena. The quantum mechanical backdrop provides a critical context for these processes, embedding uncertainty, probabilistic outcomes, and entanglement into the framework of string interactions.

One key aspect where quantum mechanics plays a pivotal role is in the arena of string vibration modes. According to string theory, each mode corresponds to a different particle, and the superposition principle of quantum mechanics enables these strings to exist simultaneously in multiple states. This not only accounts for the existence of different types of particles but also allows for the transformation of one type of particle into another through string interactions such as splitting and joining.

Probabilistic Nature and Quantum Field Theory: In traditional quantum mechanics, particles have probabilities described by the wave function's square magnitude. Similarly, string theory uses the wave function to describe probabilities but does so over the configuration space of strings rather than point particles. Herein, quantum field theory (QFT) becomes integral. QFT, which describes particles as excitations of fields, has its framework adapted in string theory to encompass strings. The fields in string theory are not fields of point particles but fields of strings, which significantly enriches the theoretical landscape by integrating an infinite array of vibrational modes.

Another pivotal area affected by quantum mechanics within string theory is in the calculation of interaction probabilities. String theory and quantum mechanics converge to provide methods using Feynman diagrams annotated for string theory. These diagrams visually depict how strings propagate through space-time and interact with each other, incorporating quantum uncertainty and

5.4. THE ROLE OF QUANTUM MECHANICS IN DESCRIBING STRING INTERACTIONS

allowing for calculations of probabilities of various string interaction outcomes.

Quantum Entanglement and Coherence: In quantum physics, entanglement is a phenomenon where particles become interlinked, and their physical properties correlate with each other even when separated by large distances. String theory extends this concept to strings, where entanglement can occur not just at the point-particle level but also involving their vibrational states and modes. Such entanglement could potentially underlie the obscure yet profound phenomena like black hole entropy and the information paradox.

Additionally, coherence—an aspect where all parts of a system exhibit quantum mechanical behavior in a unified way—is crucial in maintaining the integrity of string interactions over quantum fluctuations. Quantum coherence ensures that despite the underlying probabilistic nature and potential for decoherence through environmental interactions, certain characteristics of strings are preserved across their dynamic evolutions.

Furthermore, the application of advanced mathematical techniques such as path integrals facilitates a deeper understanding of how these strings interact within a quantum framework. Path integrals generalize the classical action path from one point to another into a sum over an infinite number of possible paths. In string theory, this concept is adapted into summing over not just paths but also topologies of the string world sheets (surfaces traced out over time by strings in space).

Integral to advancing this association between quantum mechanics and string theory is computational simulation, which often utilizes complex algorithms to explore potential scenarios and interactions based on quantum probabilities. These simulations help to illuminate how quantum mechanics shapes the behavior and interaction of strings, providing insightful views into unseen corners of theoretical physics.

The connection between quantum mechanics and string theory not only deepens our understanding of particle physics but opens new avenues in the exploration of quantum gravity, potentially unraveling new dimensions of the universe's fundamental fabric.

5.5 Quantum Entanglement and Superposition in String Theory

Quantum entanglement and superposition, fundamental aspects of quantum mechanics, permeate the fabric of string theory, presenting a comprehensive framework where these phenomena are not only preserved but also manifest with unique characteristics intrinsic to strings.

Delving first into the phenomenon of superposition in string theory, each string can theoretically exist in multiple quantum states simultaneously. This is akin to the superposition principle seen in basic quantum systems where particles like electrons or photons exist in multiple states until measured. In string theory, these states pertain to the vibrational modes of strings, which determines their properties such as mass and charge. Consider a string in state ψ_1 representing an electron and another state ψ_2 representing a neutrino. The string's total state can be expressed as a linear combination,

$$\Psi = \alpha\psi_1 + \beta\psi_2,$$

where α and β are coefficients representing the probabil-

5.5. QUANTUM ENTANGLEMENT AND SUPERPOSITION IN STRING THEORY

ity amplitudes for the string to be in either state. The key difference in string theory lies in the multidimensional aspect of these vibrations, which are spread across the additional spatial dimensions posited by the theory.

Quantum entanglement in string theory goes beyond the inherent complexity seen in quantum mechanics. When two particles are entangled, measuring a property of one instantaneously affects the property of another, regardless of the distance separating them, encapsulating the phenomenon known as "spooky action at a distance." In string theory, this entanglement can occur not just between point particles but between extended objects like strings. Two entangled strings, when split or when they interact, continue to exhibit correlated properties across large spatial dimensions of the universe. The incredible aspect here is the dimensional topology of these strings, which adds layers to how these entanglements can configure and influence each other, complicating the typical quantum mechanical view of particle interaction.

The Superposition and entanglement of strings bring profound implications for understanding complex systems and potentially for technology like quantum computing, where strings can be used to represent qubits in a computationally powerful quantum supercomputer.

To illustrate, consider a simple example depicted through a graphical representation:

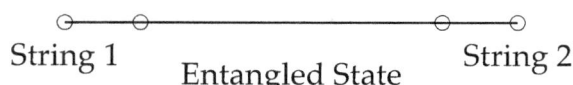

String 1   Entangled State   String 2

This diagram simplifies the abstraction, where String 1 and String 2 are entangled across a space indicated by a dotted line, symbolizing quantum entanglement.

Understanding of these entanglements and superpositions in a higher-dimensional space as per string theory potentially leads us toward insights into the deepest questions about the fundamental structure of our universe and the forces governing it, pushing the boundaries of what we perceive as locality and information transfer.

As we further explore these concepts, it becomes evident that string theory not only incorporates quantum mechanics but extends it into realms previously unachievable with conventional theories alone, promising new avenues for both theoretical insights and practical applications.

5.6 Quantum Fluctuations and String Dynamics

Delving into the essence of string dynamics, one must understand the crucial role played by quantum fluctuations within this domain. Quantum fluctuations, often envisioned as temporary changes in the amount of energy in a point in space, act as a fundamental mechanism in quantum field theory. These fluctuations are pivotal in string theory as they give rise to the dynamic characteristics of strings, contributing to their non-static, vibrant nature.

Consider a simple analogy from the classical guitar. When a guitar string vibrates, the sound produced fluctuates naturally due to minor variations in string tension, finger placement, and atmospheric conditions. Similarly, in string theory, strings undergo quantum fluctuations. However, instead of air pressure or string tension, these fluctuations are influenced by quantum probabilities. Such fluctuations introduce variations in the energy lev-

5.6. QUANTUM FLUCTUATIONS AND STRING DYNAMICS

els of strings, leading to a myriad of potential quantum states.

Given the higher-dimensional nature of strings as compared with point particles, the dimensionality adds layers of complexity to these fluctuations. A one-dimensional object fluctuating in a multi-dimensional space will have vastly different behaviors and interactions from a zero-dimensional point particle in standard quantum mechanics. This dimensional distinction gives strings their unique properties and theoretically allows them to unify all fundamental interactions of nature.

Moving deeper into the mathematics, the Polyakov action, an integral part of string theory which describes how strings propagate through space and time, provides a means to quantify these fluctuations. The action itself is expressed as an integral over the worldsheet of the string. Within the framework of the Polyakov action, quantum fluctuations are visualized through variations in the worldsheet metric, which embodies how strings stretch and interact with spacetime.

In regards to string dynamics, these quantum fluctuations are not merely perturbations but are essential for maintaining the coherence and stability of string states. The resulting dynamics from these interactions are highly nonlinear and intricate, requiring sophisticated mathematical tools for their description and analysis. The use of perturbative techniques, where interactions are treated as small deviations from a known simpler problem, often becomes inadequate. Hence, non-perturbative methods, such as those using dualities and M-theory, become indispensable.

For a more illustrative understanding, imagine plotting the potential energy landscapes originating from these quantum fluctuations on a multidimensional graph. Each point on this landscape corresponds to a possible state of a vibrating string, influenced by quantum fluctuations. These landscapes are often rugged with multiple local minima and maxima; a string could theoretically exist in any of these minima due to fluctuations pushing it from one minimum to another. This model helps encapsulate the complex dynamics and the non-intuitive behavior of strings influenced by quantum phenomena.

In essence, the behavioral panorama of strings under the influence of quantum fluctuations reveals a rich tapestry where possibility meets reality at the quantum level. The implications of this relationship are vast and potentially revolutionary, indicating paths forward not just for further theoretical developments within string theory but also for congruent fields that hinge upon the fundamental principles of quantum mechanics and quantum field theory.

As we appreciate these myriad developments, it becomes ever more apparent that the dynamic interplay between quantum fluctuations and string dynamics is not just a theoretical curiosity but a cornerstone in our quest for a unified description of nature's forces. This understanding beckons a closer inspection and deeper penetration into the wonders of string theory as it dances to the quantum tune.

5.7 Non-locality and Locality Perspectives in String Theory

At its core, string theory counters certain classical intuitions about physics, dramatically altering our under-

5.7. NON-LOCALITY AND LOCALITY PERSPECTIVES IN STRING THEORY

standing of space and time. This departure is particularly evident in the debate of non-locality versus locality in the theory.

The concepts of locality and non-locality are deeply embedded in physics. Locality asserts that objects are only directly influenced by their immediate surroundings. However, non-locality, as suggested within quantum mechanics, indicates that particles can influence one another at a distance instantaneously. String theory introduces novel perspectives on these principles.

To understand this within the framework of string theory, consider first the conventional locality in quantum field theory. It states that points separated by spacetime intervals cannot influence one another unless mediated by local fields traveling at sub-light speeds. This idea underpins the Standard Model of particle physics—but string theory demands a broader canvas.

In string theory, the fundamental constituents of reality are not zero-dimensional point particles but one-dimensional strings. These strings can stretch, merge, and split, intricately weaving through multiple dimensions. This inherent flexibility allows for phenomena that challenge conventional locality. When two strings split or merge, for instance, the effect is not confined to a single point, but distributed along the length of the string. Thus, interactions in string theory inherently integrate an extended form of locality, where the notion of point-based interaction no longer holds.

String theory also entertains a form of non-locality. Considered through the lens of dualities, a fascinating aspect of string theory, non-local connections emerge naturally. T-duality, an essential concept in string theory, shows that strings propagating in a compactified extra dimension can exhibit behaviors where the physics is identical despite different geometrical compactifications. Here, action at one point of a string seems to be mysteriously connected to another, displaying an intrinsic non-local characteristic.

Quantum entanglement provides another ground where non-locality and locality interplay fascinatingly in string theory. Typically exemplified by the EPR paradox and Bell's theorem in quantum mechanics, entanglement suggests a holistic perspective where separated quantum systems can display correlated behaviors that defy classical explanations. In string theory, the holographic principle strengthens this notion—proposing that a lower-dimensional boundary can entirely describe a higher-dimensional volume non-locally. This organically aligns with the ideas from quantum mechanics but extends them to a cosmological canvas.

In summary, string theory accommodates and extends both locality and non-locality. The strings' extended object nature requires a redefinition of locality in physical interactions, naturally incorporating non-local characteristics without contradicting the foundational principles of causality and relativistic constraints. This duality of perspectives—not merely merging but transforming foundational principles—illuminates our comprehension of the universe's fabric at its most fundamental level. It challenges us to rethink not only the nature of particles and forces but also the texture of spacetime itself.

5.8 Quantum Gravity: The Joint Venture of String Theory and Quantum Mechanics

Quantum gravity occupies a unique niche at the confluence of quantum mechanics and string theory, seeking to establish a coherent theoretical framework that explains gravity within the principles of quantum mechanics, an arena traditionally dominated by general relativity. This section delves into how these two theories converge to address one of the most elusive phenomena in physics: the quantum nature of gravitation.

The quest for a quantum theory of gravity presents one of the major challenges in modern theoretical physics. Traditional quantum mechanics allows us to describe three of the four fundamental forces—electromagnetic, strong, and weak interactions—with great accuracy. However, gravity remains notably absent from this list, as it resists formulation under the standard quantum framework provided by the theory of general relativity.

String theory introduces a novel perspective by suggesting that particles, including gravitons—which in theory mediate gravitational forces—are not zero-dimensional points, but rather one-dimensional strings. These strings vibrate at specific frequencies, and their vibrations correlate with different particles. The inclusion of these strings naturally integrates gravity by incorporating a vibrational state corresponding to a graviton. This lean towards a unified approach in string theory provides a fresh outlook on how gravity could be understood in quantum terms.

The fundamental assertion of string theory positing that

these strings can split and combine provides a mechanism for explaining how gravitational interactions might occur at the quantum level. When strings split or join, they could exchange gravitons, essentially enabling a quantum-level description of gravitational forces. Here, perhaps, lies the most significant contribution of string theory to quantum gravity—offering a mechanism that is absent in the particle-point perspective of classical quantum mechanics.

Continuing further, the impact of string theory on quantum gravity becomes even more profound when considering the behavior of strings at Planck scale distances. At these microscopic scales, traditional notions of spacetime cease to function predictably, requiring a framework like string theory that inherently includes a minimum length scale — the length of a string. This inclusion avoids the singularities typically associated with general relativity and potentially leads to a finite description of spacetime at even the tiniest scales.

One particularly promising aspect is the emergence of extra dimensions within string theory models. Quantum mechanics and general relativity do not necessitate more than four spacetime dimensions, yet string theory inherently proposes additional spatial dimensions that could fundamentally alter our understanding of gravity. These additional dimensions provide a richer framework which might integrate gravitational force more seamlessly with the other fundamental forces, thus supporting a more unified field theory.

Furthermore, string theory also contributes to quantum gravity through the theoretical prediction of phenomena such as black hole thermodynamics. Investigations have shown that under the framework of string theory, the

5.8. QUANTUM GRAVITY: THE JOINT VENTURE OF STRING THEORY AND QUANTUM MECHANICS

entropy of black holes, which marks a significant intersection with quantum mechanics, could be effectively explained. This not only supports the idea that quantum mechanics and gravity can coexist but also bridges significant gaps in our comprehensive understanding of black holes.

To visualize these theoretical contributions, consider the analogy of strings in a complex multilayered fabric. Each string's activity—splitting, merging, vibrating—introduces subtle yet profound impacts on the fabric's overall behavior, analogous to gravitons influencing spacetime's curvature at quantum scales. The concept transcends traditional disconnects between quantum physics and gravitational theory, hinting at a deeply interwoven relationship between these fundamental aspects of our universe.

This intricate collaboration between string theory and quantum mechanics in pursuit of a functional understanding of quantum gravity showcases not only the depth of theoretical challenges but also the potential for profound revelations in our foundational understanding of the universe.

This section illuminates the significance and intricacies of integrating string theory with quantum mechanics to demystify the quantum nature of gravity, employing a range of theoretical constructs that pave the way for further explorations and insights into the fabric of the cosmos. Through this joint venture, the realm of physics continues to advance toward more comprehensive theories that bridge known and unknown phenomena.

5.9 Holographic Principle: A Revolutionary Insight

The holographic principle posits a fascinating concept: all the information contained within a volume of space can be represented as encoded information on the boundary of that space. This boundary, unlike any conventional boundary, does not need to occupy a higher-dimensional space. In essence, this principle suggests that the entirety of our voluminous universe could be understood as a two-dimensional surface.

Initially conceived by physicist Gerard 't Hooft and later extended by Leonard Susskind, the holographic principle gained significant traction after Juan Maldacena's formulation of the AdS/CFT correspondence - a duality between a gravitational theory in $(d+1)$-dimensional Anti-de Sitter (AdS) space and a conformal field theory (CFT) in d dimensions. This correspondence serves not only as a concrete realization of the holographic principle but also as a bridge linking quantum mechanics and gravitational theories, paramount in string theory.

AdS/CFT Correspondence: The Gateway to String Theory and Quantum Mechanics This duality is particularly illuminative in integrating concepts from quantum mechanics into string theory. Consider string theory, which posits that point-like particles in particle physics are replaced by one-dimensional "strings." These strings can interact, bifurcate, and merge, with different modes of vibrations corresponding to different particles. The dynamics of these strings, inherently quantum mechanical, exhibit behavior that is profoundly impacted by gravitational interactions.

The AdS/CFT correspondence implies that the complex interactions among these quantum strings can be described by simpler physical theories defined at the boundary of the space, drastically simplifying computations and predictions. Here, quantum fluctuations and entanglements of strings are encoded in the boundary, which can be described by a lower-dimensional CFT.

Mathematical Representation and Predictive Power
Mathematically, this correspondence can be expressed in the form of partition functions or generating functions of both theories being equivalent, i.e., $Z_{AdS} = Z_{CFT}$. It not only offers a robust framework for describing the thermodynamics of black holes but also sheds light on quantum aspects of gravity.

Sample Representation of AdS/CFT Correspondence

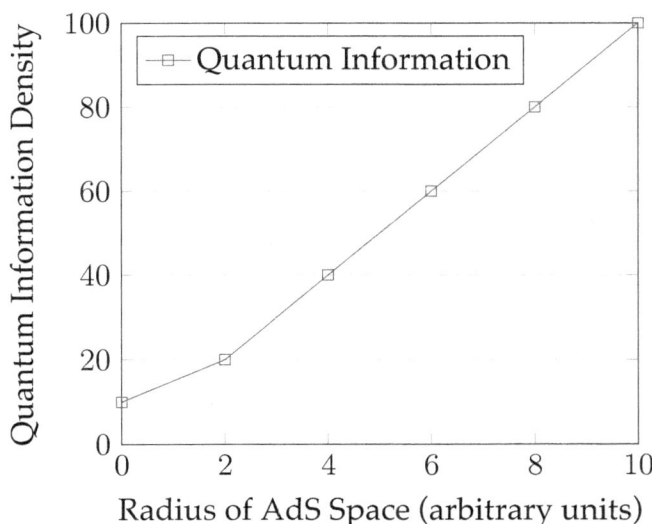

The empirically verifiable predictions of the AdS/CFT

correspondence continue to stir interest and research in both theoretical and applied physics realms. Moreover, this model provides a computational advantage to solve problems in quantum gravity by translating them into more manageable problems in a non-gravitational theory.

One of the most profound implications of the holographic principle and the AdS/CFT correspondence in string theory and its formulation is their collective potential to solve complex quantum gravitational puzzles. For example, this has significant bearings on understanding black hole entropy and informational paradoxes - areas where traditional theories of physics stumble. It introduces a new dimension (or reduces one, rather intriguingly) on how we perceive dimensions and fundamental physical laws.

These explorations ultimately pave the way for more comprehensive theories that can seamlessly integrate gravity with the quantum world - a long-standing goal in physics.

5.10 Challenges and Opportunities in Merging Quantum Mechanics with String Theory

As the discourse of integrating quantum mechanics with string theory progresses, we encounter significant challenges as well as potential opportunities that could redefine our understanding of physics. Tackling these complexities requires a nuanced comprehension of both domains and their interaction.

5.10. CHALLENGES AND OPPORTUNITIES IN MERGING QUANTUM MECHANICS WITH STRING THEORY

Challenges: Merging quantum mechanics with string theory introduces a suite of theoretical and experimental challenges. One main hurdle is the immense mathematical complexity that arises when attempting to unify these frameworks. Quantum mechanics operates with proven mathematical formalisms that govern particle behavior in the well-tested arenas of physics. However, string theory, proposing that point-like particles are replaced by one-dimensional strings, necessitates a far-reaching generalization of these existing equations.

Another major challenge arises from the lack of direct experimental evidence supporting string theory. Whereas quantum mechanics has been extensively validated through experiments, string theory's predictions often occur at energy scales inaccessible to current technology. Thus, testable predictions that can bridge quantum mechanics and string theory effectively are scarce and demand a leap in experimental techniques and technology.

Furthermore, the conceptual foundations of both theories can be occasionally at odds. For instance, locality in quantum mechanics implies non-separability, whereas string theory, with its extra dimensions and the concept of extended objects, adds layers that complicate this principle. This incongruity must be addressed to achieve a coherent theory that stands rigorous testing.

Opportunities: Despite these challenges, merging these giants of theoretical physics offers transformative opportunities. One of the most promising aspects is the potential to achieve a grand unified theory that not only incorporates the forces described by quantum field theory but also gravity, which has remained elusive within the quantum framework. This unification is expected to provide groundbreaking insights into the early universe, black holes, and quantum gravity.

Additionally, working towards this integration could propel advances in mathematical physics. Theoretical work in string theory has already led to developments in various branches of mathematics like topology and algebraic geometry. Deepening the relationship between quantum mechanics and string theory may further enrich these mathematical landscapes providing new tools for both theoretical and applied physics.

Integrating string theory with quantum mechanics also promises to enrich our conceptual understanding of space-time. String theory's depiction of a holographic universe where higher-dimensional space-time can be described by lower-dimensional boundaries could revolutionize our understanding of quantum mechanics and information theory. This opens up new avenues in quantum computing and may lead to developments in how we process and handle information on a fundamental level.

Visual Representation: To illustrate these interactions and the scale of energies at which string theory operates in comparison to observable quantum mechanical effects, consider a simple logarithmic scale.

5.10. CHALLENGES AND OPPORTUNITIES IN MERGING QUANTUM MECHANICS WITH STRING THEORY

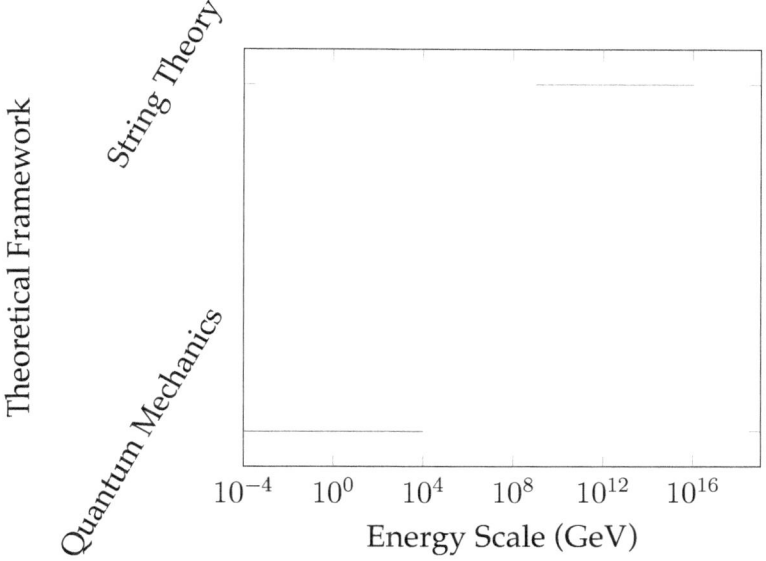

The horizontal lines represent the energy scales where each theory is primarily applicable. Quantum mechanics spans from extremely low energies up to about 10^4 GeV where traditional particle physics operates. In contrast, string theory predictions start taking over at about 10^9 GeV and above.

As we progress through the synthesis of these two pivotal areas of modern physics, it becomes essential to confront these challenges head-on while leveraging every opportunity they present to usher in new paradigms in scientific thought and technological advancement.

CHAPTER 5. QUANTUM MECHANICS AND STRING THEORY: A SYMBIOTIC RELATIONSHIP

Chapter 6

Key Experiments and Observational Evidences

This chapter focuses on the empirical aspects of string theory, outlining the key experiments and observational evidence that either supports or challenges the tenets of the theory. It details theoretical predictions of string theory that have been tested through various experimental setups including particle accelerators, gravitational wave observatories, and cosmological observations. Furthermore, it discusses the methods used for probing extra dimensions and the search for microscopic black holes, essential experiments that could potentially validate or refute aspects of string theory in the future.

6.1 The Challenge of Testing String Theory

The viability of any scientific theory fundamentally rests on its testability through empirical evidence. This is precisely the crux of the challenge when it comes to string theory, a paradigm attempting to marry the worlds of quantum physics and general relativity. The core hypothesis posits that what we perceive as particles are actually one-dimensional "strings" that vibrate at specific frequencies. However, these strings are conjectured to be on the order of the Planck length, approximately 10^{-35} meters, a scale far beyond the reach of current or near-future experimental technologies. This immense gap presents significant hurdles but also fascinating opportunities for theoretical and experimental physics.

Dimensional and Energy Constraints:

Firstly, the sheer minuteness of string scale imposes a formidable barrier. Current technological capabilities allow us to probe down to about 10^{-18} meters. To directly observe strings, we would require an energy scale around the Planck energy, 10^{19} GeV, which exceeds the capabilities of our most powerful particle accelerators by several orders of magnitude. The Large Hadron Collider (LHC), for instance, operates at a maximum of around 14 TeV, or 10^4 GeV. These limitations necessitate indirect approaches to testing string theory predictions.

Indirect Experimental Strategies:

In lieu of direct observation, physicists rely on indirect consequences of string theory that might manifest at accessible energy scales. For example, certain models of string theory predict the existence of large extra dimensions compared to the Planck scale. These could potentially be observable via high-energy particle collisions that imply gravitational effects unaccounted for by the Standard Model. Such observations would be seminal, implying a radical overhaul of our understanding of space-time dimensions.

Cosmological Footprints:

Aside from particle accelerators, cosmological signatures offer another avenue for probing the tenets of string theory. Distinct imprints in the cosmic microwave background (CMB) radiation or specific polarization patterns might provide indirect evidence of string theory dynamics in the early universe. These observations connect to string theory through theoretical models that describe the geometry and dynamics of the early universe, potentially revealing hints about the unification of forces and nature of fundamental particles.

Computational Techniques in String Theory:

Tackling the challenge from another angle, advancements in computational physics have enabled simulations and numerical calculations that explore the implications of string theory. Techniques such as lattice gauge theory and Monte Carlo simulations are used to study lower-dimensional analogues of string phenomena, which might offer insights by analogy. Computational complexity remains a bottleneck here, given that simulating even simple predictions requires substantial computational resources.

Future Directions:

Looking forward, the development of new technologies such as higher-energy particle colliders, more sensitive detectors, and advanced space telescopes could gradually erode these barriers. Furthermore, interdisciplinary approaches encompassing quantum computing and machine learning might evolve into crucial tools for cracking some of the thorniest issues in string theory. These endeavors not only aim to find direct or indirect evidence supporting string theory but also enrich our understanding of physics at both the minutest and most expansive scales.

Through a fusion of experimental intuition and theoretical innovation, physicists continue the intricate task of testing string theory. Despite substantial challenges, each step taken—be it incremental or groundbreaking—adds a valuable stitch to the fabric of our cosmic understanding, inching us closer to a unified framework of fundamental forces and matter.

6.2 Historical Experiments Influencing String Theory Development

The development of string theory has been significantly influenced by a series of key historical experiments. Each set the stage not through direct evidence of strings, but by shaping the theoretical landscape in which string theory evolved. We begin by examining the Michelson-Morley experiment, a pivotal moment in physics that challenged prevailing notions about the luminiferous ether, the medium then thought necessary for light propagation. Though devised to detect ether, the null result instead subtly pointed towards the need for new theories of space and time, eventually leading to Einstein's relativity. This non-detection of ether was critical: it suggested that not all fundamental components of nature are directly

6.2. HISTORICAL EXPERIMENTS INFLUENCING STRING THEORY DEVELOPMENT

observable, a poetic precursor to the hidden dimensions posited by string theory.

Advancing in time and complexity, the discovery of the electron by J.J. Thomson demonstrated particles could be components of atoms, previously believed indivisible. This revelation that atoms themselves were composites including smaller, subatomic entities extended into the string theory framework, wherein what were thought to be elementary particles (like quarks and electrons) are envisioned as excitations on a string.

The next substantial influence came from the gold foil experiment by Rutherford, which overturned the plum pudding model of the atom, presenting a nuclear model with a concentrated central charge. This experiment underscored the layered structure within atoms and suggested that deeper yet undiscovered layers could exist, mirroring the additional spatial dimensions and intricate vibrational modes string theory proposes.

Quantum mechanics introduced fundamentally probabilistic elements to physics, deviating sharply from the deterministic views of classical mechanics. The famous double-slit experiment elegantly demonstrated wave-particle duality, illustrating that light could exhibit properties of both waves and particles depending on experimental conditions. String theory similarly posits dualities, where objects can have dual descriptions as particles in one frame or as a wave (or string) in another.

The emergence of quantum field theory (QFT) and its triumph in unifying electromagnetism and weak nuclear forces into the electroweak theory provided a template for a unified theory of all interactions, something string theory aspires to achieve at a deeper level including gravity. These developments had laid the groundwork showing that diverse forces could have common origins—an idea string theory extends by arguing that all particles and fundamental forces arise from one single type of object: strings.

Each experiment contributed a conceptual pillar on which string theory was later built. Although none set out with string theory in mind, their outcomes indirectly guided theoretical physics towards a universe far more interconnected and dynamic than previously thought. Through this historical narrative, it becomes clear how past scientific milestones echo within the modern corridors of string theory, each advancing our approach to understanding the fabric of the cosmos.

This section illustrates how seminal experiments in the history of physics have inadvertently paved the way for string theory by shaping fundamental concepts about space, matter, and duality.

6.3 Observational Evidence for Extra Dimensions

Discovering extra dimensions is crucial for substantiating several predictions of string theory, notably the concept that spatial dimensions beyond the familiar three-dimensional space can exist. Such a discovery would propel our understanding of the universe into new territories, transforming abstract theoretical constructs into observable realities.

A key stepping stone in the search for extra dimensions comes from the study of gravitational forces. According to general relativity, gravity operates within the three di-

6.3. OBSERVATIONAL EVIDENCE FOR EXTRA DIMENSIONS

mensions of space and one of time we are accustomed to. However, string theory posits that gravity might also permeate other, unseen dimensions. The strength of gravity could appear significantly weakened in our familiar three-dimensional space because it is dispersed across these additional dimensions.

To explore this hypothesis, physicists have embarked on experiments to detect deviations from Newton's inverse-square law at short distances, which states that the force between two masses is inversely proportional to the square of the distance between them. Experiments conducted using highly sensitive torsion balances or atomic force microscopes have attempted to measure gravitational strength at scales as small as a fraction of a millimeter. While no deviations have been found to date that would suggest the presence of extra dimensions, these experiments must push technological boundaries further, refining the sensitivity and scale at which measurements are made.

Additionally, the Large Hadron Collider (LHC) has played a pivotal role in searching for extra dimensions through high-energy particle collisions. One theory suggests that if extra dimensions exist, particles called gravitons may escape into these dimensions. The escape of gravitons would manifest as missing energy and momentum in particle collision events because they would be invisible to our detectors focused on observing the familiar dimensions. Analysis of collision data looking for such losses has yet to confirm the existence of extra dimensions but continues to provide upper limits on the size and nature of these dimensions.

Cosmological observations also offer potential indirect evidence for extra dimensions. Alterations in the strength of gravitational interactions could have subtle but distinct effects on the evolution of the universe, influencing the Cosmic Microwave Background radiation, the distribution of galaxies, and other astronomical phenomena. The precise measurement of these cosmic entities and phenomena allows scientists to set constraints on theories involving extra dimensions.

Incorporating the quest for observational evidence of extra dimensions in a manner engaging to the readers involves not just sharing these facts but delving into their implications. Suppose the existence of extra dimensions is confirmed; this will not only support string theory but also drive major revisions in our comprehension of space, time, and the fundamental forces of nature. Thus, the vigil continues with an air of anticipation, each experiment an opportunity to peel back yet another layer of the universe's grand design.Pioneering techniques and relentless inquiry hold promise for what might eventually be one of the most profound discoveries in the history of science.

6.4 Methods of Detecting Vibrational Patterns of Strings

The fundamental premise of string theory posits that the most elementary particles are not point-like dots but instead tiny, vibrating strings. Each vibration corresponds to a different particle, with the mode of vibration determining the type of particle. Detecting these vibrational patterns is thus pivotal for the validation of string theory. Here we delve into the sophisticated methods proposed and utilized to detect these patterns, encapsulating their

6.4. METHODS OF DETECTING VIBRATIONAL PATTERNS OF STRINGS

principles, applications, and the challenges they behold.

One significant method is the use of high-energy particle collisions. Particle accelerators, like the Large Hadron Collider (LHC), propel subatomic particles to high energies before smashing them together. These collisions can theoretically excite the vibrational states of strings, assuming they exist at accessible energy scales. The detectors surrounding these collision points capture and analyze the resultant particles, potentially including those directly linked to string vibrations.

In these collisions, one looks for unusual signatures that ordinary particles do not exhibit. For example, jet formations following collisions could indicate new particles that might be manifestations of vibrating strings. However, differentiating these signals from the standard model noise remains a substantial challenge.

Another approach involves indirect detection via astrophysical observations. The vibrational modes of strings might influence the behavior of cosmic rays and the microwave background radiation — remnants of the Big Bang. Modifications in the expected energy spectrum of these elements can hint at deviations caused by string dynamics. Ground and space-based telescopes equipped to measure such anomalies provide vital data. Data from observatories like the Planck satellite, which maps the cosmic microwave background (CMB), are scrutinized for unusual fluctuations that might be attributable to string vibrations.

Additionally, theoretical advancements have prompted the examination of gravitational waves as a method to detect string vibrations. Gravitational wave observatories like LIGO and Virgo could potentially observe oscillations in spacetime created by vibrating strings. If string theory adjusts the texture of spacetime as it suggests, ripples in spacetime themselves might carry imprints of string vibrations, visible in the data from these observatories.

Analyzing the data from these diverse sources necessitates complex computational models. Theoretical physicists employ intricate simulations to predict possible outcomes from string vibrations and compare these predictions against experimental data. This comparison is crucial in determining whether an observed phenomenon could indeed be a manifestation of string theory or simply an artifact of more classical physics.

The technologies underpinning these methods are still in their developmental stages. Achieving the necessary sensitivity and specificity to definitively detect string vibrations poses significant technological challenges. Moreover, the energy scales at which string effects would unmistakably stand out—potentially as high as the Planck scale (about 10^{19} GeV)—are immensely beyond our current experimental reach.

Despite these challenges, the endeavor to detect vibrational patterns of strings continues to drive a significant portion of experimental physics, representing an intriguing intersection of theory and technology. As we push the boundaries of what our technologies can probe, each experiment brings us closer to understanding whether the poetic premise of string theory holds the foundational truths of our universe.

6.5 Utilizing Particle Accelerators: Exploring String Theory Predictions

Particle accelerators, such as the Large Hadron Collider (LHC) at CERN, are quintessential in the pursuit to examine the predictions of string theory. These monumental machines propel subatomic particles to nearly the speed of light before colliding them, creating an array of exotic, high-energy events ripe for analysis.

The interplay between string theory and particle accelerators can be appreciated through the specific phenomena they explore, such as the production of microscopic black holes and the search for evidence of extra dimensions. According to string theory, these aspects should be observable if the energy levels and conditions are appropriate, which accelerators are designed to achieve.

1. Microscopic Black Holes:
One of the more dramatic predictions of string theory is that particle accelerators might generate microscopic black holes. The formation of these tiny entities would be a groundbreaking confirmation of extra dimensions. The theory posits that for black holes to be generated at accelerator energies, the scale of gravity's strength must reduce significantly, which would be indicative of extra dimensions influencing gravitational force at small scales.

The process expected is quite straightforward; high-energy collisions might squeeze enough energy into a small enough volume to replicate the density needed to create a black hole. Although initial experiments at the LHC have not yet verified their existence, the ongoing adjustments in energy levels and detection sensitivity hold potential for future discoveries.

2. Extra Dimensions:

The search for extra dimensions through particle accelerators hinges on detecting energy patterns that deviate from the Standard Model of particle physics. When particles collide at high energies, as in the LHC, if there are additional dimensions compactified at scales near or within our observable reach, some energy from the collisions might "leak" into these dimensions. This leakage would appear indirectly as missing energy in the collision's aftermath, not accounted for by known physics.

Detailed analysis traces out variations in energy dispersal, allowing physicists to infer the possible presence of hidden dimensions from anomalous energy signatures.

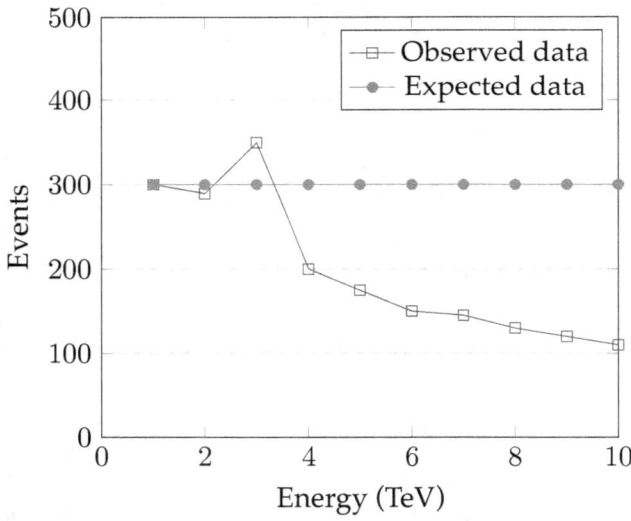

Hypothetical Missing Energy Distribution in Particle Collisions

In this visualization, discrepancies between 'Observed data' and 'Expected data' signify potential indicators of physics beyond the Standard Model, possibly pointing to extra dimensions according to string theory.

Cross-referencing these experimental observations with theoretical calculations emphasizes the interconnected fabric between theoretical physics and tangible experimental designs, marking an era where complex theories like string theory slowly unravel through advanced technological experiments. Moving ahead, particle accelerators continue to offer a window into these profound questions, continually stretching the boundaries of human knowledge about the fundamental structure of everything.

6.6 Gravitational Wave Detection and Implications for String Theory

Gravitational waves, the ripples in the fabric of spacetime, serve as a profound testament to the dynamism of the universe. These waves were predicted by Einstein's General Theory of Relativity and were first observed directly by the LIGO Scientific Collaboration in 2015. The discovery not only confirmed one of the last unresolved predictions of relativity but also opened a new observational frontier for string theory.

String theory, a theoretical framework that posits that fundamental particles are not zero-dimensional points but rather one-dimensional "strings," suggests unique signatures that could be manifested in the gravitational waves spectrum. These signatures might arise from the vibrational frequencies of strings, potentially offering a unique fingerprint indicative of higher dimensions or other exotic phenomena postulated by string theory.

Theoretical Predictions: For gravitational waves, string theory posits several novel features. One prediction is the potential modification of the standard graviton model. In string theory, gravitons could have counterparts with higher spins or might exhibit behaviors implying interaction types not predicted by the Standard Model or General Relativity.

Moreover, the creation of microscopic black holes during gravitational wave events could be indicative of extra dimensions, as these phenomena are more likely in scenarios involving more than four dimensions. Detecting such occurrences would bolster the notion of extra-dimensional spaces posited by string theory and could potentially provide a methodology to measure the size and shape of these dimensions.

Observations and Experiments: The current generation of gravitational wave detectors, including LIGO and Virgo, are designed primarily to observe and confirm phenomena predicted by General Relativity. However, concepts for new experimental setups or modifications to existing detectors are being explored to make them sensitive to the effects predicted by string theory. These modifications involve enhancing the detectors' sensitivity to higher frequency vibrations or to potential secondary effects induced by exotic phenomena like cosmic strings or additional graviton modes.

One exciting speculative notion in string theory tantalized by gravitational wave research is that cosmic strings, if they exist, would interact gravitationally with black holes or neutron stars, potentially leaving traceable imprints in the gravitational waves emitted during such interactions. Detecting these signals would provide unprecedented insights into the structure and dynamics of the early universe.

Implications: If detected, the specific signatures of strings in gravitational waves would fundamentally alter our understanding of the universe's fundamental structures. Such findings would validate several predictions of string theory, providing a much-needed experimental foundation for the theory's larger claims about the universe's makeup.

In designing future detectors and refining analysis techniques, scientists continue to incorporate theoretical insights from string theory to enhance the scope of gravitational wave astronomy. This melding of theory and observation is not just bolstering our understanding of known phenomena but is also guiding us toward the subtle whispers of the universe that could reveal the true nature of reality, as envisaged by string theory.

Crafting experiments to detect these subtle predictions of string theory through gravitational waves not only pushes forward the capabilities of our technology but also deepens our theoretical understanding, spreading light on areas that remain merely speculative today.

6.7 Cosmological Observations: Insights into Branes and Extra Dimensions

The exploration of branes and extra dimensions through cosmological observations is a fascinating frontier in modern physics, providing critical insights that challenge and expand our understanding of the universe. In this context, branes are multidimensional objects within higher-dimensional space, which according to string theory, host the fundamental strings whose vibrations manifest as the elementary particles we observe. The concept of extra dimensions extends beyond the familiar four (three spatial dimensions plus time), suggesting new realms that are tightly wrapped and hidden at scales inaccessible by current direct measurement technologies.

A pivotal cosmological probe in this arena is the study of the Cosmic Microwave Background (CMB) radiation. The CMB radiation is the thermal remnant from the Big Bang and serves as a deep-space backdrop that contains subtle imprints of the very early universe. By examining anomalies and irregularities in the CMB data, scientists can infer the presence of extra dimensions. For instance, extra dimensions could influence the way gravity propagates through the universe, affecting the CMB's temperature fluctuations. Advanced data analysis from missions like the Planck satellite has provided high-resolution maps of the CMB, where these subtle deviations indicative of higher dimensions could potentially be identified.

Another key cosmological observation relevant to string theory involves the study of dark matter and dark energy. These enigmatic components constitute about 95% of the total mass-energy content of the universe and could be clues to understanding higher-dimensional spaces. Theorists suggest that dark matter could consist of particles known as WIMPs (Weakly Interacting Massive Particles), which might originate from higher-dimensional spaces and only interact weakly with ordinary matter. The behavior and distribution of dark matter mapped in cosmic structures like galaxy clusters and gravitational lensing patterns could reveal interactions influenced by extra di-

6.7. COSMOLOGICAL OBSERVATIONS: INSIGHTS INTO BRANES AND EXTRA DIMENSIONS

mensions.

Further cosmological insights come from the study of high-energy cosmic rays. These are immensely energetic particles that travel through space and strike the Earth's atmosphere. Some models based on string theory predict that collisions between branes in higher-dimensional spaces could release fluxes of high-energy particles. If correlations between observed high-energy cosmic rays and theoretical predictions can be established, this could provide indirect evidence for the existence of branes interacting in additional dimensions.

Using these observations, physicists attempt to either verify or constrain the predictions made by string theory regarding extra dimensions and branes. Through rigorous analysis, it becomes possible to determine whether these predictions align with the empirical evidence gathered from cosmological phenomena. Each bit of data integrated into this framework not only tests the limits of current theories but also guides the development of more refined models that may one day offer a complete description of the universe's fundamental nature.

Crucially, as this endeavor progresses, it embodies a dual significance—shedding light on the uncharted territories of cosmological science while simultaneously testing the tenets of one of the most profound theoretical frameworks in modern physics. Through each observational stride, we inch closer to potentially confirming or refuting elements of string theory, perhaps even paving the way towards unveiling a new layer of reality hidden within the enigmatic dimensions predicted. As we peer deeper into these cosmic signatures, the universe might just reveal some of its most guarded secrets through the intangible whispers of branes and beyond.

6.8 Microscopic Black Holes and String Theory Tests

Microscopic black holes, also referred to as quantum black holes or mini black holes, represent a critical juncture in the experimental investigation of string theory. These theoretical phenomena are predicated on the idea that at very small scales, the gravitational force could manifest in similar ways as the other fundamental forces, potentially being unified under the framework of string theory. This section will delve deeply into the current scientific approaches and theoretical underpinnings that guide our quest to detect or infer the existence of these minute entities, which if detected, could provide an unprecedented confirmation of string theory's validity.

The concept of microscopic black holes arises from the extrapolation of general relativity and quantum mechanics, fundamentally linked to the scales described by Planck units. In scenarios of high-energy particle collisions, where energies reach or exceed the Planck energy (approximately 10^{19} GeV), the resultant density and curvature of spacetime could theoretically become so extreme that it causes a miniature black hole to form. Despite their name, these black holes would have masses on the order of the Planck mass (about 2.176×10^{-8} kilograms) but extremely tiny Schwarzschild radii.

Detection of such microscopic black holes would serve as a significant pointer to the granularity of space at quantum scales, proposed by string theory. The Large Hadron Collider (LHC), operated by CERN, is one of the few experimental setups capable of approaching the energy densities considered necessary for the formation of these phenomena. High-energy collisions produced at the LHC

6.8. MICROSCOPIC BLACK HOLES AND STRING THEORY TESTS

can potentially reproduce the conditions required, albeit conditions are constrained significantly by safety protocols ensuring that any produced black hole would pose no threat, decaying almost instantaneously via Hawking radiation.

We employ a methodological approach in theoretical physics known as perturbative string scattering amplitudes to calculate the probability of quantum black hole formation in particle collisions. Consider the following generalized expression for the scattering amplitude in string theory:

$$A(s,t) = g_s^2 \int \frac{d^2\sigma}{(2\pi)^2} |X'(\sigma_1)X'(\sigma_2)|^s |X(\sigma_1) - X(\sigma_2)|^t e^{ik\cdot X}$$

Here, g_s denotes the string coupling constant, σ parameters indicate points on the string worldsheet, and s and t channel Mandelstam variables relate to energy and momentum transfer, respectively. If this amplitude predicts a non-negligible probability for energies attainable at current or future colliders, this would provide indirect but valuable evidence supporting string theory.

Another theoretical implication of microscopic black holes within string theory relates to higher-dimensional theories such as the braneworld scenarios. In these models, our observable universe is a three-dimensional surface (brane) embedded within a higher-dimensional space. Microscopic black holes might not only form at high energies but could also provide clues about the structure of these extra dimensions.

An exciting potential comes from astrophysical signatures. Certain theoretical models predict that microscopic black holes could form in the ultra-high-energy cosmic rays striking the Earth's atmosphere. Observatories like the Pierre Auger Observatory are at the forefront of detecting such high-energy events, which could hint at microscopic black holes through an analysis of shower patterns and energy distributions.

As we progress further into high-energy physics experiments and more sensitive detections in cosmic ray observations, our understanding of whether microscopic black holes can form—and what their properties might be if they do—will significantly refine our understanding of both string theory and our universe's ultimate nature. This fascinating intersection of theoretical prediction and practical experimentation continues to push the boundaries of what we know about the cosmos and the underlying laws governing it.

6.9 Role of String Theory in Explaining Quantum Phenomena

One of the most profound challenges in modern theoretical physics has been the reconciliation of quantum mechanics and general relativity. Historically, quantum phenomena have been described successfully using quantum field theories, whereas general relativity has provided an excellent description of the gravitational force at large scales. A major contribution of string theory is its potential to unify these two fundamental but seemingly incompatible frameworks into a cohesive whole.

Quantum mechanics describes the behavior of particles at the smallest scales where phenomena such as superposition, entanglement, and uncertainty dominate. In classical quantum mechanisms like Quantum Field Theory

6.9. ROLE OF STRING THEORY IN EXPLAINING QUANTUM PHENOMENA

(QFT), particles are considered point-like, leading to various infinities or singularities when trying to integrate with general relativity. Herein lies the first pivotal role of string theory: strings replace point particles. In string theory, particles are not zero-dimensional points but rather one-dimensional strings. These strings have a characteristic length, which is typically on the order of the Planck length (1.6×10^{-35} meters), providing a natural cutoff for ultraviolet divergences that plague quantum field theories.

Let us delve deeper into the dual role strings play in unifying forces. One notable feature is how string theory accommodates multiple vibrations or modes, each corresponding to a different particle in the standard model of particle physics. This natural emergence of particles from strings effectively integrates gravity derived from the curvature of spacetime, as postulated by general relativity, with the other three fundamental forces—electromagnetic, strong, and weak nuclear forces—thereby advancing a unified framework.

Quantum Gravity and String Theory: The quest for a quantum theory of gravity finds a promising lead in string theory. Quantum gravity attempts to describe gravity according to the principles of quantum mechanics, and where previous attempts to formulate a quantum gravity theory resulted in unresolvable infinities, string theory provided a breakthrough. The graviton, a hypothetical quantum particle that mediates gravitational force, emerges naturally in string theory as a closed loop, unlike electrons or quarks that appear as open-ended strings. This suggests an intrinsic compatibility of string theory with quantum phenomena at a gravitational scale.

Now, consider the conceptual experiment involving black

holes. String theory contributes significantly to quantum phenomena through the explanation of black hole entropy and temperature - aspects historically treated separately in quantum mechanics and general relativity. Stephen Hawking's discovery that black holes radiate heat over time and gradually evaporate introduced a quantum aspect to these massive gravitational objects. String theory models these processes by expressing the microstates of black holes as states of strings and branes, providing a microscopic explanation to these macroscopic phenomena. This supports the idea that string theory effectively models quantum phenomena at all scales.

Furthermore, the holographic principle, inspired by string theory, posits that all the information contained in a volume of space can be represented as a theory on the boundary of that space. This principle has provided profound insights into quantum information theory and has implications for understanding quantum entanglement and information paradoxes.

Visualization Through Path Integrals: To represent these interactions visually, the path integral approach in string theory can be modeled. Consider the following simplified representation:

$$\int [Dx] \exp\left(\frac{i}{\hbar} S[x]\right),$$

where $S[x]$ represents the action over the path x, and Dx indicates integration over all possible string paths. This formalism illustrates how transition amplitudes between different quantum states can be conceptualized within string theory, providing a broader framework for understanding the quantum phenomena traditionally modeled through non-gravitational quantum field theories.

In essence, string theory not only proposes a theoretical model where quantum phenomena mesh smoothly with gravitational theory, but also significantly enhances our understanding of high-energy physics, cosmology, and black hole physics.

6.10 Future Experiments and Technologies to Test String Theory

Advancements in experimental physics and technology could significantly enhance our ability to test predictions derived from string theory. One of the principal challenges in testing string theory is its requirement for energy scales far beyond the capacity of current particle accelerators. Therefore, the next generation of experiments and technologies is aimed at reaching these unprecedented energies or discovering alternative methods to probe the fundamental principles of the theory.

Development of Higher Energy Particle Accelerators: Future particle accelerators, such as the proposed Future Circular Collider (FCC) and the Compact Linear Collider (CLIC), are designed to reach collision energies significantly higher than those of the Large Hadron Collider (LHC). With the LHC reaching energies up to around 13 TeV, proposals for these future colliders suggest they could achieve energies as high as 100 TeV. Such an increase would not only improve the precision of tests for existing theories like the Standard Model but could also provide the necessary conditions to explore phenomena predicted by string theory, including the production of microscopic black holes or the detection of supersymmetric particles.

Advanced Gravitational Wave Observatories: The observation of gravitational waves has opened a new avenue for astrophysical research and also holds promise for testing string theory. The next phase in gravitational wave astronomy involves the development of more sensitive detectors such as the Laser Interferometer Space Antenna (LISA) and the Einstein Telescope. These instruments aim to detect waves from a broader array of sources at greater distances. Such capabilities are crucial for investigating the cosmological imprints of branes and testing the string theory predictions regarding the early universe's evolution and potential multiple dimensions.

Search for Cosmic Strings: Cosmic strings, hypothesized one-dimensional topological defects in space-time, are predicted by some models of string theory. Detecting these objects would be a breakthrough in validating string theory. Future experiments leveraging more detailed cosmic microwave background (CMB) data and new polarization studies could potentially identify signatures characteristic of cosmic strings, such as distinct patterns in gravitational lensing.

Quantum Computing and Simulation: Quantum computers hold the potential to simulate particle interactions at an unprecedented scale and with high accuracy, providing another pathway to test aspects of string theory that are currently computationally inaccessible. Advances in quantum algorithm development and increased qubit coherence times could enable the simulation of quantum gravity and space-time dynamics as described by string theory.

Neutrino Telescopes and Dark Matter Detectors: While

6.10. FUTURE EXPERIMENTS AND TECHNOLOGIES TO TEST STRING THEORY

neutrino observatories and dark matter detection experiments primarily seek to elucidate aspects of these elusive components of our universe, they may inadvertently provide insights into higher dimensions or string-theoretic particles. Enhancements in detector sensitivity and coverage could allow these observatories to capture rare events that current models of particle physics cannot explain.

In pursuing answers through these advanced technologies and experimental setups, scientists are not merely testing string theory but also exploring the boundaries of human knowledge about the fundamental constituents of reality. As these initiatives progress, they may reveal not only whether string theory holds any truth but also deeper insights into the universe that could reshape our understanding of all physical phenomena.

CHAPTER 6. KEY EXPERIMENTS AND OBSERVATIONAL EVIDENCES

Chapter 7

String Theory and the Nature of Reality: Space, Time, and Matter

This chapter delves into the profound implications of string theory on our understanding of the fundamental aspects of reality, specifically focusing on concepts of space, time, and matter. It discusses how string theory redefines these elements, proposing new frameworks that contrast with classical and relativistic views. The narrative explains the roles of strings and branes in shaping the structure of the universe and revisits the traditional views of time and space in light of string theory's propositions. Additionally, the chapter investigates how string theory's perspective on matter provides a more unified understanding of the universe's fundamental constituents.

CHAPTER 7. STRING THEORY AND THE NATURE OF REALITY: SPACE, TIME, AND MATTER

7.1 Redefining Space and Time through String Theory

String theory offers a revolutionary perspective on the fundamental constructs of space and time, challenging and extending the classical and relativistic frameworks that have long formed the basis of our understanding of the universe. In traditional physics, space and time are treated as separate entities; space providing a static backdrop for events, and time progressing uniformly independent of the observer's motion or position. However, the advent of Einstein's theory of relativity melded space and time into a single four-dimensional continuum known as spacetime, where the geometry could warp and curve in response to mass and energy.

String theory takes this concept even further by suggesting that space and time may not be fundamental constructs but emergent phenomena arising from more basic entities: strings and branes. These are not zero-dimensional points but rather one-dimensional loops and higher-dimensional membranes whose vibrations and interactions give rise to the properties and dimensions of space and time as we perceive them.

The radical notion posited by string theory that space and time are emergent implies that at the most fundamental level, the universe is composed of tiny strings whose vibrational states can generate not only particles of matter but the very fabric of spacetime itself. Moreover, string theory introduces additional dimensions – beyond the familiar three of space and one of time – which are compactified and curled up at scales so small that they remain undetectable with current technology.

7.1. REDEFINING SPACE AND TIME THROUGH STRING THEORY

One way to conceptualize these extra dimensions is through mathematical models which illustrate how strings vibrating in multiple dimensions might manifest as different particles in lower dimensions. For instance, a string vibrating in one way might appear to us as an electron, while another mode of vibration might be perceptible as a photon. The nature of these vibrations, and how they translate to effects in lower dimensions, is pivotal in explaining various fundamental phenomena such as particle masses and the forces between them.

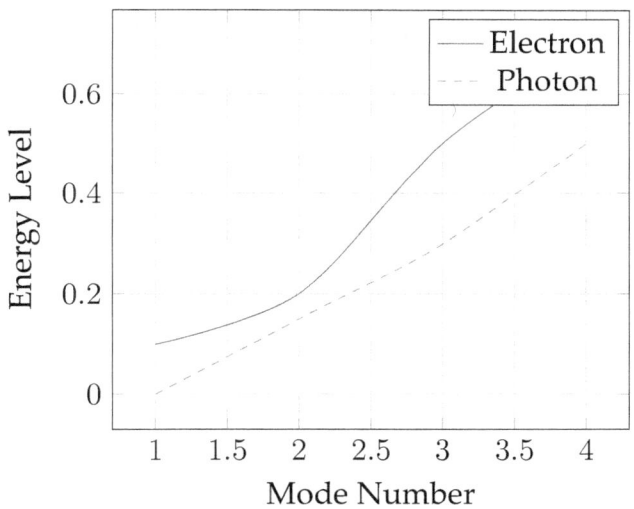

Hypothetical Representation of Vibrational States

This framework not only redefines our understanding of space and time but also underpins revolutionary approaches to solving persistent problems in physics, such as the unification of general relativity and quantum mechanics. In string theory, the smooth spacetime manifold breaks down at the Planck scale, revealing a foamy, quantum mechanical nature that might eventually provide a way to reconcile these foundational theories.

Furthermore, the implications for cosmology are profound. The big bang, traditionally understood as the origin of spacetime from a singular point, might instead be seen as a transition in the state of an ever-existing set of strings. This offers a radically different view of the universe's beginning, suggesting a cyclical or ever-iterative model of creation and dissolution.

Exploring these groundbreaking ideas deepens not only our understanding of what space and time are but fundamentally challenges how we perceive reality and our place within it. Through the lens of string theory, we begin to see that space and time may not be the rigid, unchanging plains we once considered them but are instead dynamic, responsive fabrics woven from the cosmic symphony of strings.

7.2 The Concept of Matter in String Theory

Delving into the concept of matter within string theory reveals a universe fundamentally different from that depicted by classical physics. In traditional frameworks, matter is composed of particles delineated by specific points in space; however, string theory introduces a paradigm where these point particles are replaced with one-dimensional objects known as strings. These strings vibrate at distinct frequencies, and these vibrations determine the type of elementary particles they represent, such as quarks or electrons.

This notion is revolutionary because it implies that the particles making up atoms, and consequently all matter, are not fixed "building blocks" but dynamic entities with

7.2. THE CONCEPT OF MATTER IN STRING THEORY

fluctuating properties determined by vibrational states. To visualize this, imagine a guitar string. When plucked, it can produce different musical notes depending on factors like tension and length. Similarly, in string theory, strings can oscillate in multiple modes, and each mode corresponds to a different particle in nature.

Mathematically, the vibrational state of a string is described by the harmonic oscillation equations, which are integral in defining the string's energy and resultant mass. Using the Planck constant $\hbar$, and the speed of light c, these properties integrate into the fabric of space-time, governed by Einstein's theory of relativity. This integration illustrates how string theory seamlessly fuses the micro (quantum) and macro (relativistic) aspects of the universe.

$$E = \sqrt{p^2c^2 + m^2c^4}$$

Equation 1, representing the relativistic energy-momentum relation, shows how energy (E) depends on momentum (p) and mass (m). In string theory, these parameters are dictated by how strings oscillate, thus linking string vibration directly to fundamental properties like mass and charge.

Moreover, string theory suggests that matter's properties emerge from the geometry and the interaction dynamics of these strings. For instance, when two strings split or merge—an interaction mimicking particle collisions in quantum physics—their vibrational states alter, leading to changes in particle types and characteristics. This feature accounts for the rich variety of particles observed in nature and activities within particle accelerators.

In essence, string theory presents a universe where all

CHAPTER 7. STRING THEORY AND THE NATURE OF REALITY: SPACE, TIME, AND MATTER

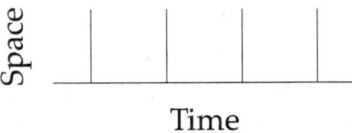

Figure 7.1: Simplified representation of string interactions over time

matter is fundamentally interconnected through the vibrations and interactions of strings across multiple dimensions, some of which extend beyond our conventional three-dimensional perception. Exploring further into these dimensions offers potential explanations for uncommon physical phenomena such as particle types that have yet to fit neatly within standard models, or the dark matter problem—matter that influences gravitational forces but does not emit any electromagnetic radiation.

Embracing string theory allows us to visualize not only what matter consists of but also the extraordinary and beautifully complex symphony it performs at the cosmic scale—a symphony where the notes, represented by fundamental particles, come together to comprise the observable universe.

As we continue to investigate and refine this model, the very fabric of our understanding of reality undergoes a profound transformation, urging us to revisit fundamental questions about what matter is and how it fundamentally shapes our universe. Through this lens, string theory not only redefines matter; it challenges and extends our very conception of reality.

7.3 How String Theory Modifies Our Understanding of Gravity

String theory introduces a radically different perspective on gravity, diverging significantly from Newtonian physics and General Relativity. Central to this modification is the very premise of string theory, where the most fundamental constituents of the universe are not point particles, but rather tiny, vibrating strings. The oscillations of these strings in multiple dimensions generate the diverse particles and forces observed in our universe, including gravity. This contrasts sharply with classical notions, where gravity is a force caused by the mass of objects warping space-time around them according to Einstein's theory of General Relativity.

In classical physics, gravity is treated as a force that acts at a distance between two masses. Einstein's theory refined this concept by proposing that mass curves the fabric of space-time, and gravity is the effect of this curvature. Objects move along the paths dictated by this geometry. String theory, however, posits that these phenomena arise from the vibrations of strings. Each mode of vibration corresponds to a different particle. Remarkably, one such vibrational mode of a string corresponds to the graviton, a hypothetical quantum particle believed to mediate gravitational forces. Unlike other theories of gravity, string theory naturally integrates gravity with quantum mechanics.

The modification of gravity in string theory can be illustrated by examining its quantum properties. Traditional approaches have failed to unify gravity with quantum mechanics, leading to significant theoretical challenges. In string theory, the scale at which quantum effects become significant is the Planck length, approximately 1.6×10^{-35} meters. At this tiny scale, the structure of spacetime becomes influenced by quantum fluctuations. Here, string theory's portrayal of gravitons not as point particles but as extended objects helps in mitigating problems like infinities arising in quantum field theories due to point particle interactions. The finite size of strings smoothens out the sharp distinctions seen at quantum points, leading to a theory less plagued by infinities.

Moreover, string theory posits extra spatial dimensions which could provide a mechanism through which gravity weakens at short distances. According to this view, our observable universe is embedded in a higher-dimensional space. Some dimensions are compactified and curled up at scales so small that they escape detection. This setting changes the gravitational force law at small distances and might help in explaining why gravity is markedly weaker than other fundamental forces, a longstanding puzzle in physics.

In practical terms, studying how string theory affects our understanding of gravity could potentially resolve several issues unresolved by General Relativity and quantum mechanics. These include understanding the quantum aspects of black holes, such as Hawking radiation or the information paradox. Furthermore, it might provide a framework for devising experiments or interpretations regarding dark matter and dark energy—concepts that are critical in cosmology but are poorly understood within the confines of General Relativity alone.

Comprehending gravity through string theory suggests a universe where the fabric of reality consists of tiny strings whose harmonious vibrations orchestrate the cosmos's symphony, including the gravitational interactions.

This is more than a theoretical curiosity; it represents a profound shift in our paradigm for understanding every force in the universe. In exploring these concepts, we stand at the brink of potentially unifying all fundamental forces under a single comprehensive framework, illuminating dark corners of our understanding and possibly revealing new aspects of reality that are yet to be imagined.

7.4 The Relationship between Vibrating Strings and Fundamental Particles

Deepening our exploration into the core concepts of string theory, it is crucial to understand the profound connection between the theory's quintessential entities—vibrating strings—and the fundamental particles which constitute all known matter. String theory posits that what we perceive as particles are in fact minuscule, vibrating strands of energy, each oscillating at distinct frequencies. These strings, unlike point particles of traditional particle physics, possess length but no other dimensions.

To immerse into this relationship, let's first consider the different types of strings proposed by the theory: open strings, which have two distinct endpoints, and closed strings, forming a complete loop. Depending on the mode of vibration and the nature of the string (open or closed), these strings manifest as different particles. For instance, the photon, a carrier of electromagnetic force, is represented as a particular oscillation mode of an open string. In contrast, gravitons—mediators of gravitational force—are hypothesized to arise from closed string vibrations.

Each string can vibrate in numerous ways. Each mode of vibration corresponds to a different mass and charge for the particle that the string represents. This principle helps unify seemingly unrelated particles. For example, an electron and its antiparticle, the positron, can be understood as the same type of string vibrating in opposite ways. This duality and versatility suggest a harmonious underlying principle where diversity in particles emerges from different vibrational states of the same fundamental objects.

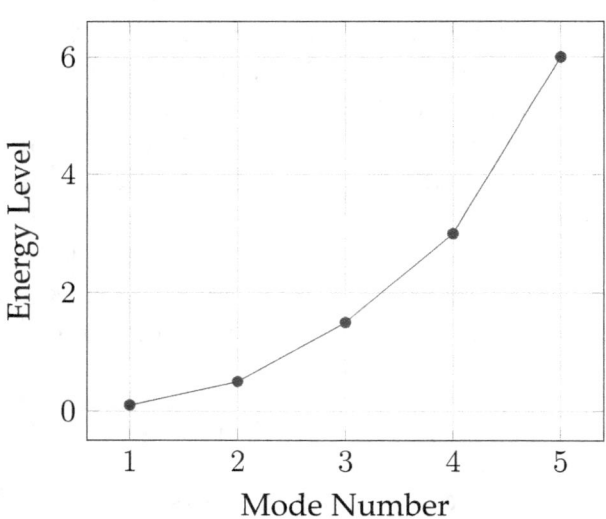

The multiple vibration modes translate into a multitude of possible particle configurations. To calculate the properties like mass and charge associated with different vibration states, string theorists employ advanced mathematical frameworks. Compactification, or the process by which extra string dimensions curl up, plays a pivotal role in shaping these properties. The manner and size of these curled dimensions determine the characteristics imparted to the particles through string vibrations.

String theory further introduces fascinating hypotheses about the formation of these particles. When strings split or join, new particles are formed—a scenario portraying particle interactions at a subatomic level. Such events underline the dynamic and interconnected nature of matter. This approach contrasts starkly with the traditional, static view of particles and introduces a concept where change and transformation are constant at the fundamental particle level.

Crucially, the relationship between vibrating strings and fundamental particles leads to a potentially unified theory of all forces and matter constituents. Known as the Theory of Everything, this ambitious perspective suggests that all physical phenomena originating from varied forces (gravity, electromagnetism, weak and strong nuclear forces) can be explained via different states and configurations of vibrating strings. This coherence promises not just to redefine our understanding of the universe's elementary particles, but also challenges our entire perception of reality at its most basal level.

This holistic view emerges not just as a theoretical feast but as a paradigm inviting deeper insights into the universe's most esoteric aspects. Through the stringent framework provided by string theory, the dance of vibrating strings gently pulls the curtain aside to reveal a universe far more interconnected and dynamic than discerned by classical physics, setting a profound stage where the microcosm and macrocosm interlace seamlessly.

7.5 Space-Time Fabric and Membranes in String Theory

Space-time and membranes, or branes as they are often called, serve as central concepts in string theory that significantly extend our understanding of the universe's fabric. To appreciate the innovation that string theory introduces, we must first understand these concepts not as separate entities but as interconnected components of a higher-dimensional universe.

String theory posits that what we perceive as particles are actually tiny, vibrating strings. These strings can be open or closed; open strings have two separate endpoints, while closed strings form a complete loop. The modes of vibration of these strings determine the types of particles they manifest as, including their mass and charge. However, the truly revolutionary aspect of string theory comes into play with the introduction of higher-dimensional branes.

Branes can be conceptualized as multi-dimensional membranes where strings can attach themselves. In string theory, branes can exist in various dimensions, from 1-dimensional strings (1D) up to 9-dimensional hypersurfaces (9D) in the 10-dimensional space that theories such as superstring theory propose. Fundamentally, these branes are not just passive actors; they dynamically interact with the strings and influence the overall topology

7.5. SPACE-TIME FABRIC AND MEMBRANES IN STRING THEORY

of space-time.

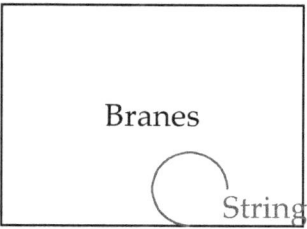

In this simplified diagram, a string (blue) is shown interacting with a brane (rectangle). The string can vibrate, loop, or even stretch from one point to another on the brane, and in higher-dimensional scenarios, it could wrap around or across multiple branes.

The fabric of space-time itself, according to string theory, isn't a fixed backdrop but a dynamic collection of intertwining branes of varying dimensions. The interactions among various branes and strings generate the physical phenomena we observe. For instance, gravity is believed to be the result of closed strings propagating between branes, which can manifest even in dimensions that aren't directly perceivable.

Another critical aspect introduced by brane theory is the concept of multi-universes or parallel universes. Branes can move and collide in the higher-dimensional space, leading to potentially observable effects such as the Big Bang, previously attributed solely to singularities. These collisions of branes can create or annihilate universes, introducing a cyclical nature to cosmology contrary to the classic linear progression.

Through intricate mathematical formulations and theoretical constructs, branes and the space-time fabric as depicted in string theory offer unique insights into particle physics, cosmology, and the fundamental structure of everything around us. This transformed framework provides a more holistic view that could potentially unlock answers to some of the most profound questions about our universe's origin, structure, and ultimate fate.

7.6 Emergent Phenomena: Space-Time from a Network of Strings

One of the most revolutionary proposals in string theory is the concept of space-time emerging from a network of vibrating strings. This idea challenges the classical notion that space and time are fundamental continua that provide a backdrop for physical phenomena. Instead, string theory suggests that space-time itself is composed of discrete, interconnected components—seemingly ethereal strands of energy. This section elaborates on how this radical shift extends our understanding of the universe's fabric.

In traditional physics, space and time are represented as continuous and unchangeable entities woven into the very fabric of reality. String theory, however, posits that at the Planck scale (approximately 10^{-35} meters), the smoothness of space-time breaks down into a frothy, dynamic mesh of strings. Each string in this network can be envisioned as a one-dimensional line oscillating in multiple dimensions.

The dynamics of these strings are governed by the principles of quantum mechanics, which implies that their states are defined by probability waves. Their symphonic vibrations give rise to the particles that form matter and convey forces. Importantly, the way these strings interact,

7.6. EMERGENT PHENOMENA: SPACE-TIME FROM A NETWORK OF STRINGS

join, and split under the dictates of string theory generates the tapestry that we perceive as space and time.

To delve deeper into this concept, let's visualize the universe as a complex orchestration where each string represents a different musical note. The interplay of these notes (strings) creates harmonies (particles) and melodies (forces), contributing to the overall composition (the universe). This analogy helps demystify how space and time could emerge from such interactions.

Simulations in string theory often use models like lattice string networks where nodes represent string endpoints while edges manifest as strings themselves. This discretized model offers computational insights into how space-time could dynamically emerge from string interactions. To illustrate this, consider a regular lattice structure of strings interconnected to form a complex three-dimensional grid. As these strings vibrate, they affect their neighbors; an intense vibration in one part of the network can influence the overall structure's topology, potentially leading to a curvature of the network that mimics gravitational effects in general relativity.

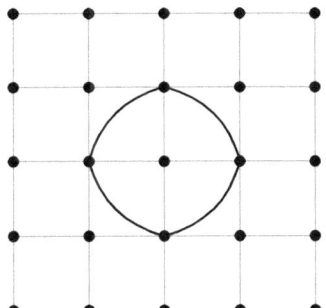

This diagram can signify a basic four-node component of a larger string network lattice where vibrations in each string segment stimulate curvatures indicating local "geometric deformities". These deformities could correspond to gravitational fields or distortions in space-time itself.

From here, the discussion extends to physical observations. Recent theoretical advancements, alongside progress in observational astronomy, suggest that these emerging phenomena might not just be theoretical constructs but observable realities. For instance, the investigation into gravitational waves and their interaction with the cosmic background could provide empirical validation for the emergent space-time theory of string networks.

In this light, string theory is more than just a theoretical framework; it provides a novel lens through which we see space and time not as passive stages but as dynamic actants in their own right - continuously influenced and fashioned by the underlying strings' symphony. The ramifications for understanding the cosmos's deeper reality are immense. As we further decode the mysteries of string theory, our perception of reality continues to evolve, suggesting that what we consider as foundational and immutable might well be beautifully transient and emergent.

7.7 Implications of Extra Dimensions on Reality Perception

One of the most intriguing aspects of string theory is its requirement for more than the three spatial dimensions we experience in our everyday life. According to string theory, there are actually ten dimensions: nine spatial dimensions and one temporal dimension. This theoretical framework catalyzes novel interpretations and per-

7.7. IMPLICATIONS OF EXTRA DIMENSIONS ON REALITY PERCEPTION

ceptions of reality which differ significantly from conventional physical theories. Let us delve into how these extra dimensions might affect our understanding and perception of the universe.

To begin, envisage a universe where dimensions beyond our familiar three are not just theoretical constructs but tangible elements of reality. In this framework, entities exist that move along these higher dimensions, just as we move in our three-dimensional space. However, because the additional dimensions are compactified - curled up at scales much smaller than atoms - they elude direct detection by current scientific instruments. Despite their elusive nature, these dimensions hold the potential to unify all the forces of nature, offering a comprehensive understanding that classical mechanics and even quantum physics could not fully provide.

The implications of extra dimensions extend beyond theoretical completeness. One concrete example is the impact on gravitational forces. In models with extra dimensions, gravity can propagate into these dimensions, leading to observable effects in the four-dimensional spacetime we inhabit. This possibly explains why gravity is markedly weaker compared to other fundamental forces; it is diluted as it spreads out through more dimensions.

Further, the concept of particles in string theory varies from the point-like particles of quantum mechanics. Particles in this enriched multidimensional context are conceptualized as modes of vibration of strings moving through spacetime, including the extra dimensions. This fundamentally alters our perception of matter and fields—each particle's properties, such as mass and charge, correlate with different vibrational patterns of the strings, influenced by the geometry of the extra dimensions.

To illustrate, consider a simple vibration model depicted as follows:

$$v(x) = A\sin(kx - \omega t + \phi)$$

Where A is the amplitude, k is the wave number, ω is the angular frequency, and ϕ is the phase. In string theory, each mode of vibration could correspond to a different elementary particle in our four-dimensional spacetime perspective. The extra dimensions might modify parameters such as k and ω, thereby altering the observable characteristics of the particles.

Moreover, the geometry and size of these extra dimensions directly affect phenomena such as charge quantization and the values of fundamental physical constants. Considering different geometrical configurations or different sizes of these extra dimensions might lead to different physical laws or variations in physical constants, thus offering potential explanations for why our universe seems to be finely tuned for life.

Another remarkable aspect lies in probing black holes. By studying the interaction of black holes with extra dimensions, researchers can uncover clues regarding the fundamental structure of spacetime itself, potentially observing effects like radiation leakage into other dimensions, which could validate string theory's predictions.

This intangible array of extra spatial dimensions offered by string theory enriches our understanding of the universe's fabric considerably. By exploring these dimensions, we unravel more about the fundamental aspects of nature that are otherwise inaccessible through traditional three-dimensional lenses. As our tools and techniques evolve, so too will our ability to investigate these extraordinary facets of reality, ultimately advancing our mastery over the universe's deepest secrets.

7.8 Time Dilation and Length Contraction: Relativity Meets String Theory

Understanding the traditional notions of time dilation and length contraction through the lens of string theory provides a novel pathway to explore the confluence between relativity and quantum physics. While Einstein's theory of relativity beautifully describes the behavior of objects in high-speed motion and significant gravitational fields, string theory introduces a multi-dimensional universe where these concepts may manifest differently.

In special relativity, time dilation refers to the phenomenon where time, as measured by a clock moving at a significant fraction of the speed of light, appears to pass more slowly when observed from a stationary reference frame. Standard equations governing this behavior model time dilation as a function of velocity, denoted by

$$\Delta t' = \frac{\Delta t}{\sqrt{1 - v^2/c^2}}$$

, where $\Delta t'$ is the time interval observed in the moving clock, Δt is the time interval in the stationary observer's frame, v is the velocity of the moving clock relative to the observer, and c is the speed of light.

Similarly, length contraction posits that the length of an object moving at relativistic speeds will appear shorter along the direction of motion from the point of view of a stationary observer. The relationship is quantified by

$$L' = L\sqrt{1 - v^2/c^2}$$

, where L' is the contracted length, L is the proper length when at rest relative to the observer, and v and c have their usual meanings.

When these concepts are transposed into the framework of string theory, intriguing complexities arise. Aside from its familiar ten or more spatial dimensions, string theory suggests that every point-like particle is actually a one-dimensional string whose modes of vibration underlie what we perceive as particle characteristics.

Incorporating the added dimensions and the stringy nature of matter modifies how relativity operates. For instance, length contraction in string theory must consider not just the object in motion but also the vibrational state of the strings comprising it. The contracted length might then be influenced by interactions between strings, which can depend on additional compactified dimensions that are beyond direct observation.

To illustrate, consider a string vibrating within extra spatial dimensions. The detected length contraction from our four-dimensional perspective might include contributions from these hidden dimensions, altering the simple Lorentz-FitzGerald formula. A revised metric for spacetime, which accounts for these extra dimensions and their unique geometry, could be represented by a modified tensor equation that expands on Einstein's spacetime curvature model.

Moreover, in string theory, time dilation might involve more than just the relative velocities of observers and objects. As strings move and interact across various dimensions, their vibrational patterns—which dictate mass and energy—could influence the passage of time perceived by an observer equipped only to detect changes in four-dimensional spacetime.

The implication of this theory extends to how we measure and understand gravitational fields. Since string theory posits gravity as a manifestation of strings propagating through hidden dimensions, both time dilation and length contraction might occur in novel contexts that transcend classical interpretations.

As we trace the movements and interactions of these fundamentally small constituents across dimensions both large and perceivable, and tiny and hidden, we glimpse a universe far more interconnected and dynamic than previously imagined. This synthesis of relativity and string theory not only enhances our understanding but pushes us to rethink the very fabric of reality, considering that what we observe might just be shadows cast by a richer, more intricate cosmic dance of strings.

7.9 Quantum Coherence and Decoherence in String Theory

The phenomena of quantum coherence and decoherence are pivotal in understanding the quantum behaviors of subatomic particles, which include the fundamental strings posited by string theory. Quantum coherence pertains to the wave-like nature of particles where their phases are correlated and remain in a fixed relationship. This coherence is a core principle that enables phenomena such as interference patterns and is instrumental in the operational success of devices ranging from simple lasers to complex quantum computers.

In string theory, coherence assumes a fascinating role as each string's vibration mode contributes uniquely to the coherent properties of particle-like manifestations. Consider a scenario in which multiple strings interact in a multi-dimensional space known as the Calabi-Yau space. These interactions, and the subsequent modifications in their vibrational patterns, illustrate how coherence is preserved or altered. Each mode of vibration effectively represents a different particle, and the coherence between these modes can manifest macroscopic quantum phenomena in the universe.

Decoherence, on the other hand, describes the loss of this quantum coherence, usually when a quantum system interacts with its environment in a manner that is not coherently controlled - a process that entangles the system states with the environment and leads to classical behavior. In string theory, this can be akin to strings interacting with surrounding cosmic fabric or other strings, losing their coherent phase relationship and thus transitioning from a quantum to a more classical state. This feature of decoherence is essential for understanding how classical properties emerge from a fundamentally quantum world governed by string theory.

Mathematically, this transition can be modeled using density matrices. Assume a simple model where a string state, denoted by $|\psi\rangle$, interacts with an external environment. The system's initial state can be represented by a density matrix $\rho = |\psi\rangle\langle\psi|$. As interactions occur, the system evolves into a mixed state represented by $\rho = \sum_i p_i |\psi_i\rangle\langle\psi_i|$, where $|\psi_i\rangle$ are the possible states post-interaction, and p_i their respective probabilities.

Furthermore, the presence of extra dimensions in string theory adds layers of complexity to the coherence and decoherence processes. The accessibility of additional spatial dimensions can facilitate novel modes of interaction among strings, potentially leading to distinctive patterns of coherence and decoherence that are not observable in a merely three-dimensional view.

To visualize how dimensions impact coherence, imagine plotting the probability densities of string states within these extra dimensions using a graph. Each axis represents a dimension while the spread of the plot indicates the degree of coherence: narrower plots signify higher coherence. Such visualizations can dramatically simplify the understanding of complex multi-dimensional interactions.

Thus, analyzing coherence and decoherence in string theory not only aids in comprehensively understanding quantum mechanics but also underscores the novel ways in which these concepts manifest in higher dimensional spaces, providing keen insights into the very structure of reality. As we chart these territories, the inherent interplay of string theory's predictions with observable phenomenological effects continues to challenge and enrich our conceptual framework of the universe.

7.10 Exploring the Multiverse: Realities Beyond our Own

String theory, with its unparalleled capacity to unify the laws of the physical universe, takes the concept of the multiverse from science fiction to a scientifically plausible framework. It posits not just a universe but potentially an infinite number of universes, collectively known as the "multiverse". Each universe within this multiverse can have different laws of physics, dimensions, and constants. This concept reshapes our understanding of our own reality and opens up a myriad of possibilities for the fundamental nature of existence.

The idea of the multiverse emerges naturally from various facets of string theory. The theory's requirement of extra spatial dimensions—beyond the three spatial dimensions familiar to us—means that these additional dimensions must be compactified or curled up in complex configurations. Each possible mode of compactification can lead to a distinct universe, setting the stage for a landscape of possible universes. This is often visualized using Calabi-Yau manifolds, intricate six-dimensional shapes that could define unique properties for space and time in different universes.

Calabi-Yau manifold visualization

Figure 7.2: Schematic illustration of a Calabi-Yau manifold used in string theory.

Furthermore, the vibration frequencies of strings, which determine particle types and properties according to string theory, could vary from one universe to another in the multiverse scenario. If strings vibrate differently, their corresponding universes would have distinctive

7.10. EXPLORING THE MULTIVERSE: REALITIES BEYOND OUR OWN

sets of physical laws and constants. For example, what we regard as the electron in our universe may not exist or may be entirely different with unique properties in another universe.

The anthropic principle plays a critical role in multiverse considerations, particularly in explaining the fine-tuning of constants in our universe. In an ensemble of universes, each with different laws and constants, it is probable that only a few would have the right conditions to support life. Therefore, our presence in this universe could simply be because it is one among many where conditions happened to be right for life as we know it.

Universe Characteristics	Implication for Life Existence
Exact physical constants	Permits complex chemistry and stable energy sources required for life
Alternate physical constants	Could lead to a universe with no stable atoms, no chemical reactions, or extreme instability

Testing the multiverse theory remains a major challenge. Indirect evidence might come from cosmological phenomena that cannot be easily explained by our existing theories of physics. For instance, certain features of the cosmic microwave background radiation could potentially hint at interactions with other universes.

In pondering the multiverse and its broad implications, it is as though we are standing at the shoreline of an infinitely vast cosmic ocean. With each theoretical wave, our understanding ebbs and flows, navigating through spectacular possibilities in an endeavor to grasp the nature of our reality and perhaps, realities beyond our own.

CHAPTER 7. STRING THEORY AND THE NATURE OF REALITY: SPACE, TIME, AND MATTER

The elegance of string theory not only stretches the fabric of space and time but also weaves into it the profound narrative of multiple universes, challenging us to rethink our place not just in our universe, but in the vast assembly that may exist beyond.

Chapter 8

Unification and The Grand Design: Linking Forces and Particles

This chapter analyzes string theory's ambitious endeavor to unify all fundamental forces and particles under a single theoretical framework. It discusses how string theory attempts to merge the electromagnetic, weak, and strong forces with gravity—a quest that has challenged physicists for decades. The sections cover the role of supersymmetry in balancing equations and the implications of string theory's approach to particle physics, particularly in how it reinterprets the nature and interaction of fundamental particles. Also, it examines the experimental and theoretical challenges faced in validating these unification proposals, with a focus on predictions and their correspondences with physical observations.

CHAPTER 8. UNIFICATION AND THE GRAND DESIGN: LINKING FORCES AND PARTICLES

8.1 The Quest for Unification in Physics

Unification in physics represents a profound intellectual endeavor aimed at discovering a single underlying theory that can describe all physical phenomena. The concept stretches back to Isaac Newton's unification of terrestrial and celestial mechanics, which suggested that the same natural laws applied both on Earth and in the heavens. This idea of unifying disparate forces or interactions under a single theoretical umbrella underpins much of modern physics and sets the stage for string theory's ambitious goals.

In the 19th century, James Clerk Maxwell accomplished one of the first great unifications in physics by showing that electricity and magnetism were two aspects of a single force—electromagnetism. This revelation not only altered our understanding of these forces but also laid the groundwork for the later development of electromagnetic wave theories, including the understanding of light as an electromagnetic wave.

Entering the 20th century, efforts to delve deeper into unification became intertwined with the development of quantum mechanics and later, quantum field theory. After identifying three fundamental forces—electromagnetic, weak nuclear, and strong nuclear forces—physicists proposed the Standard Model, which successfully described the interactions mediated by gauge bosons of the aforementioned forces, albeit without including gravity.

Here, Albert Einstein's attempt to unify gravity with electromagnetism, through his unified field theory, though unsuccessful, epitomized the intense desire within the physics community to marry the theories of the large (general relativity) and the small (quantum mechanics). Einstein's vision was far ahead of its time, but it set a foundation that encouraged further research into the realm of unification.

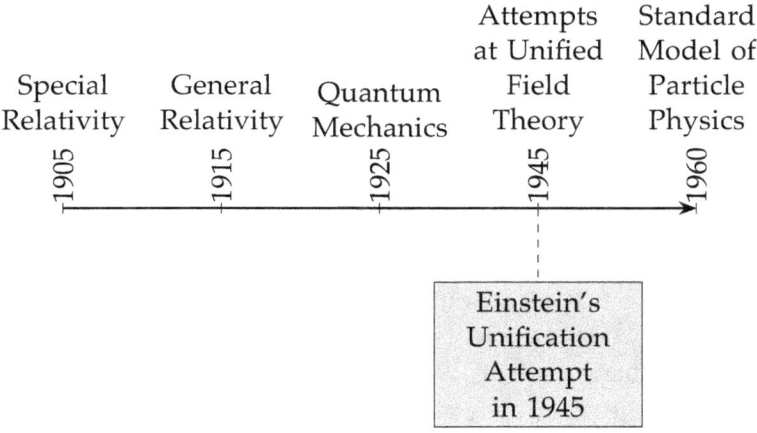

Figure 8.1: Timeline of major unification milestones in physics, highlighting key developments from Einstein's theories to modern particle physics.

The quest continued with the electroweak theory in the mid-20th century, proposed by Sheldon Glashow, Abdus Salam, and Steven Weinberg. This theory combined the electromagnetic force and the weak nuclear forces into the electroweak interaction, marking a significant milestone. This success further fueled aspirations to include the strong force, leading to the formulation of the grand unified theories (GUTs). GUTs propose that at high energy levels—as existed shortly after the Big Bang—the electromagnetic, weak, and strong nuclear forces were merged into a single force.

CHAPTER 8. UNIFICATION AND THE GRAND DESIGN: LINKING FORCES AND PARTICLES

Force	Associated Theory	Unified Theory
Electromagnetic	Maxwell's Equations	Electroweak Theory
Weak Nuclear	Flavour Dynamics	Electroweak Theory
Strong Nuclear	Quantum Chromodynamics	Grand Unified Theory
Gravity	General Relativity	String Theory

Table 8.1: Overview of force-specific theories and their unification efforts

Despite these monumental strides, gravity remained elusive from these unification efforts. General relativity, our best description of gravity, stubbornly resisted incorporation into the quantum framework established by the Standard Model. Consequently, string theory emerged as a promising candidate proposing not just to unify all fundamental forces—including gravity—but also suggesting that fundamental particles are not point-like dots but rather tiny, vibrating strings whose modes of vibration correspond to different particles.

Discovering a unified theory remains perhaps the most tantalizing quest in theoretical physics. It invites us to imagine a day when we can explain vastly complex and diverse phenomena with beautifully simple principles. As we delve deeper into string theory in this chapter, bear in mind this narrative of unification—not only as a historical pursuit but as a vibrant, ongoing dialogue among today's physicists. By weaving together these threads from history with pioneering concepts from string theory, we get closer to understanding the very fabric of the cosmos.

8.2 String Theory's Approach to Unifying Forces

String Theory proposes a breathtakingly elegant solution to one of physics' most profound puzzles: the unification of nature's fundamental forces. Traditionally, these forces are described by different theoretical frameworks; gravity is modeled by general relativity, while electromagnetism, along with the strong and weak nuclear forces, are described by quantum mechanics. String Theory, by re-envisioning these forces not as ideal points but rather as one-dimensional "strings", each vibrating at specific frequencies, attempts a revolutionary integration.

Central to understanding String Theory's approach is the concept of vibrating strings. Each string's mode of vibration corresponds to a different particle, wherein particles are not point-like dots but stretched out lines. The repercussions of such a view are profound. It permits particles to exhibit behaviors akin to both particles and waves, as observed in quantum mechanics. Moreover, the vibrational state of these strings consolidates our understanding of particle types — a groundbreaking shift from particles as unique entities to different resonant states of the same fundamental object.

A pivotal aspect in this theoretical construct is the introduction of supersymmetry. Supersymmetry, or SUSY, essentially posits that each particle has a superpartner particle differing in spin by half a unit. This concept has been instrumental in solving several theoretical inconsistencies in physics, particularly those pertaining to the hierarchy problem in particle physics. In the framework of String Theory, supersymmetry facilitates the unification of the forces by smoothing out the quantum inconsistencies that arise when trying to blend gravity with the other three forces. Consequently, String Theory posits itself as a candidate Theory of Everything (ToE) by providing a common foundation for all forces.

Moreover, String Theory also embraces more than the usual four dimensions we experience daily. The existence of extra dimensions, beyond our observable three spatial plus one temporal dimension, is crucial. These additional dimensions are typically compactified on a scale much smaller than what can be detected by current instruments. In String Theory, these compact dimensions could be tightly wound into shapes known as Calabi-Yau manifolds. The geometric characteristics of these manifolds determine how strings vibrate and interact, which in turn affects the very nature of fundamental forces and particles as perceived in our familiar four-dimensional universe.

Another significant contribution of String Theory toward unification relates to gravitational force. Gravity has been notoriously challenging to quantize due to its inherent infinitesimal properties and unyielding nature at singularities like black holes. String Theory proposes that strings at these singularities can have modes of vibration that manifest gravity at quantum scales, thus integrating gravity smoothly with the quantum framework of the other forces.

The theory's appeal is not just theoretical elegance; it extends to potentially observable implications, such as the slightly modified predictions for black hole properties and the unification scale energy levels that could be tested in particle accelerators. String Theory thereby offers testable predictions which, if verified, would lend formidable support to its unifying claims. Moreover, it suggests the possible existence of new symmetries at high energies, which could further inform our understanding of the universe.

Though densely packed with mathematical rigor, String Theory's premise that all complexity arises from simplicity provides a powerful and beautiful explanation couched in the elegance of strings: our universe woven into a vast and intricate tapestry of vibrating strings, underlying the observable complexities of particle interactions and fundamental forces.

8.3 Electromagnetism and Weak Force: Unification through Strings

Electromagnetism and the weak nuclear force, two of the four fundamental forces of nature, were historically considered disparate entities until their unification within the framework of the electroweak theory in the 1960s through the work of Sheldon Glashow, Abdus Salam, and Steven Weinberg. In the context of string theory, this unification takes on new dimensions—literally and figuratively.

String theory proposes that what we observe as different particles are actually different vibrational states of microscopic strings. The magnitude of the impact of string theory becomes more apparent when it is applied to unify forces. For instance, in conventional quantum field theories, the force-carrying particles (bosons) and matter particles (fermions) are fundamentally distinct. However, string theory introduces a framework where these distinctions blur, as both bosons and fermions arise from strings, differing only in their vibrational modes.

The unification of electromagnetism and the weak force within string theory proceeds from understanding how string vibrations can manifest as these different forces. For electromagnetism, the photon is identified as one such vibration mode of a string. In parallel, for the weak force, the W and Z bosons correspond to different vibrational states. The compelling aspect of string theory is that it encompasses these particles in a single framework whereas the Standard Model of particle physics treats them separately until symmetries are specialized under certain energy conditions.

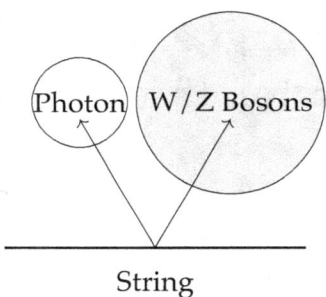

This model encapsulates how different forces can be seen as emerging from the same foundational entity—strings—which oscillate in various modes.

Moreover, in string theory, the merging of electromagnetism and the weak force is not seen merely as a feature of low-energy physics but as a necessity in understanding the full cosmological model. This includes delving into higher-dimensional spaces where these strings not only vibrate but also wrap around compactified dimensions, influencing charge, mass, and force mediator characteristics. Here lies a radical departure from traditional physics: in string theory, the characteristics of elementary particles, and thus forces, are dictated by the topology and geometry of compact dimensions unseen in our everyday four-dimensional spacetime.

To explore these concepts further, consider the role of extra dimensions in manipulating the characteristics of force-carrying particles:

Compact Dimension	Impact on Particle Physics
Shape of Compactification	Modifies Mass and Charge
Size of Compact Dimension	Influences Force Strength
Number of Dimensions	Determines Type of Symmetry

Through such a table, it is straightforward to envisage how string theory not only unifies understanding of different forces but also proposes a whole structured influence of higher dimensions on fundamental particles that represent these forces.

Thus, while the merging of electromagnetism and the weak force under the string theory umbrella profoundly echoes earlier twentieth-century unifications, it also carves out paths for potentially ground-breaking insights into the nature of everything from tiny elementary particles to the vast expanses of cosmic fabrics.

8.4 Strong Force Integration in the Context of String Theory

The integration of the strong force, one of the four fundamental forces of nature, into string theory's framework offers some of the most compelling yet intricate aspects discussed in theoretical physics. Understanding the strong force within string theory goes beyond merely cataloging particles or interactions; it involves delving into how these strings configure, vibrate, and interact at unimaginably tiny scales.

The strong force, primarily responsible for holding together the nuclei of atoms by binding protons and neutrons, is mediated by particles known as gluons. In the Standard Model of particle physics, gluons are massless gauge bosons that do not interact with the Higgs boson, unlike their fellow gauge bosons that mediate the weak force. Herein, string theory proposes a unification by suggesting that both gluons and quarks—the fundamental constituents of protons and neutrons—are different manifestations of open strings with specific vibrational modes.

To integrate the strong force in string theory, one must navigate through an elaborate landscape of higher-dimensional frameworks. One such framework is the type IIA/B string theories which incorporate D-branes—hypothetical membranes where open strings can attach. The emergence of D-branes has provided physicists with essential tools for studying non-perturbative aspects of string theory, which are crucial for understanding the strong force dynamics. Gluons can be visualized as open strings connecting these D-branes, and their interactions correspond to the merging and splitting of these strings.

This model not only accounts for the characteristics of gluons but also aligns well with their confinement in hadrons, a phenomenon inadequately addressed by the Standard Model. In string theory, the notion that quarks are connected by gluonic strings provides a geometric and topological representation of the confinement mechanism. As quarks move apart, the string elongates, and due to the string's tension, it becomes energetically favorable for the string to break forming a new quark-

8.4. STRONG FORCE INTEGRATION IN THE CONTEXT OF STRING THEORY

antiquark pair, thus explicating why quarks are never found in isolation.

Moreover, advancements in the mathematical tools necessary for string theory have led to more precise descriptions of the strong force. Techniques such as dual resonance models and holographic principles, particularly the Anti-de Sitter space/conformal field theory (AdS/CFT) correspondence, enrich our understanding by framing quantum chromodynamics (QCD)—the theory of the strong interaction in terms of a gravitational theory in higher dimensions. In this correspondence, QCD is imagined as a boundary to a higher-dimensional space, allowing for a description where gravity (via string theory) and strong force interactions are dual to each other. This duality opens up novel methods for calculating the scattering amplitudes of gluons which are notoriously difficult to handle via conventional perturbative QCD.

These theoretical advancements depict the strong force not simply as a nuclear glue but as a dynamic and geometric fabric woven from the fundamental strings. This deeper insight offers remarkable predictions and offers a fertile ground for exploring potential links between the cosmos's smallest scales and its largest forces.

Understanding and applying string theory's approach to encapsulating the strong force is a sophisticated narrative of modern physics. It does not merely represent a mathematical curiosity but embodies a broad and unified vision that might one day lead to a complete depiction of nature's forces and constituents, all woven into the expansive tapestry of the universe.

8.5 Incorporating Gravity: The Ultimate Challenge

String theory's ambitious quest to provide a comprehensive framework for all fundamental forces encounters its sternest test when it addresses gravity. Universally acknowledged as the weakest yet the most pervasive of all the forces, gravity according to general relativity is a manifestation of the curvature of spacetime caused by mass and energy. The challenge, then, is immense: string theory must reformulate gravity not as a geometric property of spacetime but as a vibrational mode of strings.

The primary tool for this formidable task is the incorporation of closed strings. Unlike open strings, which have endpoints and are generally associated with the gauge bosons that mediate the electromagnetic, weak, and strong forces, closed strings form continuous loops and have no endpoints. These closed strings are pivotal because their vibrational patterns include modes that can be interpreted as gravitons—the hypothetical quantum particles believed to mediate gravitational forces.

One of the fascinating aspects of string theory's approach to gravity is its inherent need for extra spatial dimensions. While our observed universe is manifestly four-dimensional (three spatial dimensions plus time), string theory posits that additional dimensions are required for the consistent definition of string interactions. In these extra dimensions, which are compactified and generally invisible on macroscopic scales, the strings vibrate in manners that, impressively, recapitulate general relativity's predictions on small scales.

Consider the following mathematical formulation that

8.5. INCORPORATING GRAVITY: THE ULTIMATE CHALLENGE

arises in this scenario. The action for a single closed string in a curved spacetime can be represented as:

$$S = -T \int d\tau d\sigma \sqrt{-h} h^{ab} g_{\mu\nu} \partial_a X^\mu \partial_b X^\nu,$$

where T denotes the string tension, h^{ab} is the induced metric on the worldsheet of the string, and $g_{\mu\nu}$ is the spacetime metric which connects with general relativity. The integral is taken over the string's worldsheet parameterized by σ and τ, with X^μ representing the coordinates of the string in spacetime. The graviton modes appear when we consider small fluctuations in the spacetime metric around a flat metric, $g_{\mu\nu} = \eta_{\mu\nu} + h_{\mu\nu}$ where $h_{\mu\nu}$ represents the graviton.

This equation underpins one of string theory's significant contributions to theoretical physics—the insight that not only can all particles be interpreted as different vibrational modes of fundamental strings, but also that these strings offer a route to unify all forces including gravity by varying only the modes of oscillation.

Critically, this approach has led to insights into quantum gravity—a domain where both quantum mechanics and general relativity concurrently influence physical behavior, such as near black holes or during the very early moments of the Big Bang. These scenes, where conventional physics break down, showcase the potential of string theory to deliver a consistent framework for quantum gravity.

Irrespective of its beauty and potential insights, incorporating gravity into string theory remains a venture fraught with experimental challenges. Due to the incredibly weak nature of gravity compared to other fundamental forces and the correspondingly minuscule size of the extra dimensions proposed by string theory, testing these predictions experimentally remains daunting. Particle accelerators capable of probing these energy scales have yet to be conceived, let alone built. Consequently, much of this discourse remains theoretical, awaiting future technologies or innovative methodologies capable of peering into these exceedingly fine scales.

The significance of integrating gravity extends beyond conceptual elegance; it offers hope for profound understanding and answers to fundamental questions about the universe's inception and ultimate fate. Through advancements in string theory, understanding the interplay between quantum mechanics and general relativity could not only redefine our comprehension of the cosmos but might also pave the way for revolutionary technologies in quantum computing and beyond.

In weaving together the vast tapestry that encompasses all known physical phenomena, the inclusion of gravity stands as perhaps the most formidable challenge yet for string theory. This integration not only tests the limits of contemporary physics but also poises us on the brink of potentially revolutionary breakthroughs which could transform our understanding of everything from black holes to the fabric of reality itself.

8.6 Supersymmetry: Balancing the Particle Universe

Supersymmetry, often abbreviated as SUSY, is a profound concept in theoretical physics that enhances the standard model of particle physics by proposing a partner particle, or sparticle, for every particle in the standard model.

8.6. SUPERSYMMETRY: BALANCING THE PARTICLE UNIVERSE

Central to the appeal of supersymmetry in string theory is its potential to address several challenging puzzles in contemporary physics, such as the hierarchy problem and the nature of dark matter. Furthermore, SUSY plays a crucial role in enabling the smooth incorporation of gravity with the other three fundamental forces—electromagnetic, weak, and strong forces—thus offering an elegant pathway to their unification.

In essence, supersymmetry introduces a symmetry between two fundamental classes of particles: fermions, which constitute matter like electrons and quarks, and bosons, which mediate forces such as photons and gluons. According to supersymmetry, for each fermion, there is a corresponding boson and vice versa. For instance, the electron, a fermion, would have a supersymmetric partner known as the selectron, a boson. Similarly, the gluon, which is a boson, would have a corresponding fermionic superpartner termed the gluino.

The symmetry proposed by SUSY is not apparent in the observable world under everyday conditions, as these superpartners have not yet been detected. This leads to the understanding that if supersymmetry exists, it must be a broken symmetry so that the sparticles are much heavier than their ordinary counterparts. The exact mechanism of this symmetry breaking remains an active area of research but is essential for explaining why such particles have not been observed experimentally.

One of the key motivations for supersymmetry in the context of string theory is its requirement for mathematical consistency when attempting to include gravity. String theory posits that all particles are manifestations of one-dimensional "strings" vibrating at different frequencies. Incorporating supersymmetry enhances the mathematical structure of string theory, allowing for a more stable and coherent description of the universe at this fundamental level.

In addition, supersymmetry greatly influences the string theory landscape by predicting a multitude of possible vacuum states. This multiplicity results from different ways in which supersymmetry can be broken, leading to a rich structure of possible universes with varying physical laws. Such considerations are central in string theory's pursuit to describe our particular universe among a vast array of possibilities.

Moreover, SUSY contributes to stabilizing the mass of the Higgs boson. Without supersymmetry, the mass of the Higgs boson would receive corrections from quantum effects that could make it dramatically heavier—an inconsistency with experimental observations. Supersymmetry elegantly resolves this by ensuring these corrections nearly cancel out due to contributions from superpartner loops, thereby restraining the Higgs mass within observed limits.

Let us delve deeper into how supersymmetry impacts our understanding using a simple illustration. Consider the visualization of the potential energy landscape of the universe, represented by ◯. In this model, different fields correspond to different dimensions in this landscape, and the shape depicts how the universe evolves. With SUSY, not only does this landscape gain additional dimensions due to the inclusion of sparticles, but the shapes—representing potential energy wells—become symmetrical. This symmetry imposes additional restrictions and thus refines our predictions of particle behavior.

The exploration for evidence of supersymmetry continues fervently with experiments at large colliders such as the Large Hadron Collider (LHC). Physicists worldwide eagerly await possible signals that might indicate the existence of these elusive superpartners. Discovering sparticles would not only solidify our understanding of supersymmetry but also provide direct insights into higher dimensional spaces predicted by string theory and other aspects of grand unification.

Supersymmetry thereby stands at the frontier, underpinning much of the theoretical groundwork in the quest for a unified theory of everything—gently nudging our current understanding while hinting at profound mysteries waiting to be unraveled in the grand tapestry of the cosmos.

8.7 From Particles to Strings: Transitioning the Fundamentals

Delving deep into the fabric of the cosmos, string theory proposes a strikingly elegant idea: at the most fundamental level, everything in the universe is made up of tiny, vibrating strings. This revolutionary concept suggests a shift away from considering particles as point-like entities, towards viewing them as one-dimensional strings whose modes of vibration determine the types of particles they represent. This thought paradigm not only enriches the understanding of particle physics but also opens new avenues for unifying all forces and matter.

The traditional particle physics framework, based on point-like particles, confronts several theoretical and empirical challenges as it attempts to describe interactions at incredibly small scales. Intriguing issues such as the hierarchy problem or the unification of gravity with other fundamental forces manifest limitations of the conventional models. String theory elegantly bypasses these complications by introducing strings as the basic constituents of matter and force carriers. When particles are reinterpreted as strings, their apparent point-like interactions in space are actually manifestations of strings' vibrational states.

Each type of vibration corresponds to a different particle under the Standard Model of particle physics. For instance, an electron in string theory is not a point but rather a string vibrating at a particular frequency. This reimagining has profound implications: since strings can naturally exist in higher-dimensional spaces and can interact in ways that particles cannot, they facilitate a unified framework that includes gravity naturally. The additional degrees of freedom due to the string's extended nature allow for a more cohesive integration of quantum mechanics and general relativity, something point-like particles have struggled with.

Applying these concepts, consider the visualization of fermions and bosons in string theory. Fermions, which make up matter, and bosons, which mediate interactions like electromagnetic forces, are both modeled as strings. Their interactions involve the splitting and joining of strings, which graphically can be illustrated using Feynman-like diagrams but adapted for string theory.

Consider the following basic example rendered illustrating string interaction; replace traditional particle interaction points with strings splitting or joining:

8.7. FROM PARTICLES TO STRINGS: TRANSITIONING THE FUNDAMENTALS

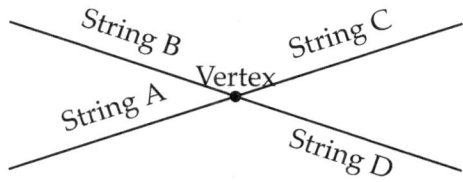

This diagram illustrates a basic interaction where two strings (A and B) merge at a vertex to form two new strings (C and D). Each string's vibrational state—its mode of vibration—determines what type of particle it is perceived as in our three-dimensional space.

Further, string theory incorporates extra dimensions which allow strings to not only vibrate but also to wind around these dimensions in complex patterns. Thus, understanding these patterns and their stability conditions necessitates a more profound comprehension of the topological and geometrical properties of the additional dimensions string theory proposes.

Approaching the unification of forces within such a paradigm involves examining how these extra dimensions influence string dynamics and, subsequently, particle interactions in our observable universe. This demands not only theoretical rigor but innovative mathematical tools and experimental approaches to probe dimensions beyond our familiar three spatial and one temporal dimensions.

As we continue to explore these ideas, it becomes clear that transitioning from a particle-centric view of the universe to a string-based theory does not just alter our theoretical frameworks; it profoundly reshapes our understanding of the universe's very essence.

8.8 Predictions of String Theory on Particle Interactions

String theory, a framework that extends beyond the Standard Model of particle physics, proposes that the familiar point-like particles are actually one-dimensional "strings" vibrating at specific frequencies. These vibrations not only determine the type of particle but also dictate how particles interact with one another. This section delves into the profound implications this has on predicting particle interactions, addressing both the theoretical underpinnings and the transformative potentials of these predictions.

Firstly, string theory introduces a new perspective on interaction mechanisms. In conventional quantum field theory, interactions are depicted by the exchange of gauge bosons, such as photons in electromagnetism. However, in string theory, interactions are the consequence of strings splitting and joining. The point of interaction in field theory, depicted as a vertex in Feynman diagrams, is replaced in string theory by a smoother scenario where strings merge or divide. This approach not only simplifies calculations by eliminating many of the infinities inherent in field theory calculations but also provides a more unified view of forces and matter.

Importantly, string theory predicts the existence of gravitons, hypothesized as closed-loop strings whose vibrations mediate gravitational forces. This is monumental because gravitons integrate gravity into the quantum framework, an achievement not accommodated by the Standard Model. The theoretical existence of gravitons suggests that string theory is capable of describing all known fundamental forces in a single framework, lead-

8.8. PREDICTIONS OF STRING THEORY ON PARTICLE INTERACTIONS

ing towards a true unification of forces.

Moreover, string theory allows for the existence of higher-dimensional interactions which are not possible in the traditional quantum physics model. These interactions involve strings propagating through extra spatial dimensions, which although compactified and not directly observable, contribute significantly to the phenomenology of particle interactions. This emerges through new possibilities for symmetry breaking and mass generation mechanisms, leading to predictions of new particles and corresponding fields.

The theoretical predictions of string theory also extend to supersymmetry - a principle that predicts every particle known has a superpartner with differing spin characteristics. A direct implication of this is that the interactions between particles are mirrored by interactions between their superpartners, potentially observable at high-energy scales. This has significant implications for not only the understanding of particle physics but also for cosmology, as these superpartners could form part of the dark matter.

String theory further predicts modifications to the scattering amplitudes of particles. Traditional perturbative techniques, used in quantum field theory, become cumbersome at higher energies due to increasing terms and complexity. String theory modifies this approach by describing scattering amplitudes as functions of the worldsheet topology formed by interacting strings. This method inherently simplifies calculations at high energies and has led to the discovery of duality symmetries among different physics theories, which suggest that seemingly complex interactions can be understood from simpler, equivalent scenarios.

In considering experimental tests, predictions from string theory are immensely challenging to verify due to the energy scales involved. Current particle accelerators cannot reach energies high enough to directly observe effects like extra dimensions or superpartners. Indirect evidence such as deviations from the Standard Model predictions in existing or future collider experiments might provide necessary clues.

In essence, string theory's predictions on particle interactions push the boundaries of current understanding and invite a reevaluation of fundamental physical principles. These insights not only deepen our comprehension of the universe but also invite innovative methods to probe the very fabric of reality.

8.9 Experimental Signs of Unification from Particle Accelerators

Particle accelerators, the colossal machines that probe the fundamental constituents of matter, play a critical role in testing the predictions of string theory. These massive structures accelerate particles to near-light speeds before colliding them. The debris from these collisions provides invaluable clues about the particles and forces at play at the most fundamental level. Central to string theory is its promise to unify all fundamental forces of nature. Consequently, accelerators are pivotal in seeking experimental evidence to support or refute this unification.

The Large Hadron Collider (LHC) at CERN, in particular, has been at the forefront of this exploration. One of the significant anticipations from string theory is the existence of supersymmetric particles – partners to the

8.9. EXPERIMENTAL SIGNS OF UNIFICATION FROM PARTICLE ACCELERATORS

known particles but with differing spins. Supersymmetry is essential for extending the Standard Model into a unified framework and is proposed to help stabilize the vast energy scales involved in unification. The LHC experiments, such as ATLAS and CMS, are designed to detect such particles. While direct evidence remains elusive, the LHC continues to push the boundaries on the mass limits where these particles might exist.

In addition to supersymmetry, string theory also suggests the possibility of detecting tiny, compactified extra dimensions through phenomena like graviton resonance or micro black holes at energy scales accessible in particle accelerators. The detection of these would revolutionize our understanding of space-time and provide concrete support for string theory. Experiments at the LHC have searched for signatures of extra dimensions through various decay patterns and missing energy measurements, though results have so far affirmed the absence of such dimensions at probed scales.

Nevertheless, it's not just the LHC contributing to this search. Tevatron, although now ceased operations, provided critical insights into the mass of the top quark and W boson, offering stringent tests for the consistency of the Standard Model and providing indirect support to theoretical frameworks like string theory that extend beyond it. Smaller-scale accelerators also contribute details on particle interactions that are pivotal in fine-tuning string theory's predictions and in exploring its proposed mechanisms like coupling constants and symmetry breakings.

To illustrate, let us consider a simplified theoretical prediction of string theory often investigated at accelerators – the energy dependency of force coupling constants. String theory proposes that at higher energies, typical of those produced in accelerators, the strengths of fundamental forces might converge. Experiments studying the running of these constants at high energies are crucial. They test whether electromagnetic, weak, and strong forces exhibit behaviours that could hint at unification under extreme conditions, as predicted.

Given the complexity and indirectness of testing string theory predictions, particle accelerators also rely on advanced data analysis techniques and theoretical models to interpret collision outcomes. Advanced computational simulations based on lattice gauge theory and Monte Carlo methods help in predicting what might be seen in an experiment, including rare events that could be signatures of new physics.

The engagement between theoretical predictions and experimental testing in particle accelerators forms a symbiotic relationship wherein each informs and refines the other's scope and directional focus. This iterative interaction is subtle yet powerful, as it steadily marshals our theoretical understanding closer to verifiable or refutable evidence, inching towards a grand unification or possibly redefining our pursuit thereof.

8.10 The Role of Extra Dimensions in Force Unification

Extra dimensions play a pivotal role in the context of string theory, providing a framework that allows for the unification of all fundamental forces, including gravity. String theory posits that, beyond the familiar three spatial dimensions and one time dimension, additional dimensions exist that are compactified and hidden from ev-

8.10. THE ROLE OF EXTRA DIMENSIONS IN FORCE UNIFICATION

eryday experience. The conjecture is that these extra dimensions facilitate the unification process by allowing for more complex vibrations of strings, which correspond to different particles and forces.

Consider the notion that each elementary particle is composed of tiny, vibrating strings, and these strings' modes of vibration determine the type of particle and its interactions. Now, if the universe has more than four dimensions, these strings can vibrate in ways that are not possible in a universe with only three spatial dimensions. It's these additional modes of vibration that give rise to the unique properties of fundamental forces as observed in nature.

To understand this better, we use a simple mathematical model where these extra dimensions are shaped like a Calabi-Yau manifold. Think of a Calabi-Yau space as akin to a multi-dimensional doughnut with a very complex topology. The specific shape and size of this manifold dictate how strings can vibrate on this compact space and thus determine the types of particles and forces in our four-dimensional universe.

Moreover, these extra dimensions are crucial in resolving discrepancies between how gravity operates and how other fundamental forces behave. In string theory, gravity emerges naturally as a mode of string vibration just like other particle types, but its apparent weakness in four-dimensional space can be explained through the dilution of gravitonic strings' vibrations into higher dimensional spaces.

Let's illustrate this with a simple analogy. Imagine blowing air through a narrow straw; the air's force feels concentrated and strong through the small opening. Now think about blowing air in an open hall; despite exerting the same amount of force, it dissipates quickly and feels much weaker. Similarly, in the context of extra dimensions, gravity might be 'spreading out' into these additional dimensions, making it appear far weaker than other forces that are confined to our familiar four dimensions.

For further clarity, here's a visual representation using a theoretical model:

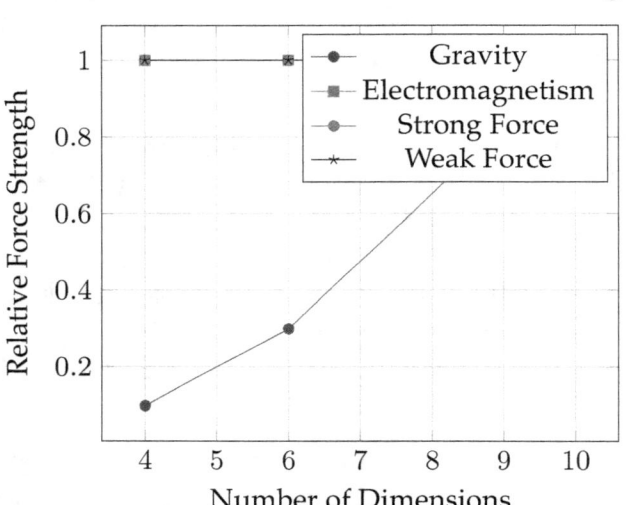

Effect of Extra Dimensions on Force Strength

What is perhaps most fascinating is that each type of force and particle is a reflection of the vibrational states afforded by these extra dimensions. Thus, by studying the number and nature of these hidden dimensions, scientists can potentially unlock a deeper understanding of the universe's fundamental forces and structure.

This intricate interplay between multiple dimensions and the fundamental particles of the universe not only opens up new avenues for theoretical physics but also sets a

8.10. THE ROLE OF EXTRA DIMENSIONS IN FORCE UNIFICATION

guidepost for future experimental innovations that could verify the existence of these extra dimensions.

CHAPTER 8. UNIFICATION AND THE GRAND DESIGN: LINKING FORCES AND PARTICLES

Chapter 9

Challenges and Critiques of String Theory

This chapter addresses the various challenges and critiques that string theory has faced since its inception. It elaborates on critical issues such as the theory's testability, the vast number of solutions in the string theory landscape, and its reliance on higher dimensions, which some argue makes the theory unfeasible. Additionally, the chapter discusses the broader scientific and philosophical implications of these critiques, including the debate over whether string theory can be considered a true scientific theory if it lacks empirical verification. The narrative also covers the responses from the string theory community and the adjustments and advancements that have been proposed to address these significant criticisms.

9.1 Overview of Major Criticisms of String Theory

String theory, as an ambitious theoretical framework in modern physics, aims to unify all fundamental interactions by conceptualizing particles not as zero-dimensional points but as one-dimensional "strings." Despite its elegance and potential, string theory has encountered a spectrum of criticisms, each addressing various facets from its scientific validity to its philosophical robustness.

One of the most poignant critiques of string theory is its current lack of testability. Fundamentally, a theory in science gains credibility through experimental verification or falsification. However, string theory operates at energy scales so high and at distances so minute that current technology falls short of directly testing its predictions. This has led to a view among some in the scientific community that string theory, as it stands, verges more on metaphysical speculation than empirical science.

Linked to the issue of testability is the critique concerning the multitude of possible solutions offered by string theory. The theory's landscape is believed to contain about 10^{500} distinct solutions, each corresponding to a different set of physical laws potentially describing a unique universe. This immense diversity, while rich in possibilities, paradoxically makes it challenging to derive definitive predictions about our own universe, thus complicating efforts to corroborate the theory experimentally.

Additionally, string theory's reliance on higher dimensions — typically requiring 10 or 11 spacetime dimensions for mathematical consistency — has been a double-edged sword. While higher dimensions provide the necessary mathematical arena for uniting gravity with quantum mechanics, they also introduce complexities and conceptual hurdles that have no observable correlates in our familiar four-dimensional worldview.

Critics also point to the absence of unique predictions as a significant limitation. For a theory to be scientifically revolutionary, it must provide unique, verifiable predictions that can be tested empirically. So far, string theory has struggled to offer predictions that are both unique to its framework and accessible through current experimental methodologies.

Lastly, competing theories like loop quantum gravity propose different approaches to the problem of quantum gravity, challenging string theory's position as the sole contender for a unified theory. These alternative frameworks sometimes offer predictions that, while also currently elusive, require less speculative groundwork regarding extra dimensions and string-based entities.

Despite these critiques, string theory continues to evolve, with researchers exploring new mathematical tools and conceptual adjustments. It stands as a profound but controversial endeavour in theoretical physics, embodying both the aspirations and the intricate challenges of seeking a unified description of nature's forces.

9.2 The Problem of Testability and Falsifiability

String theory, poised elegantly at the nexus of theoretical physics and cosmology, presents a framework that extends beyond the Standard Model of particle physics with its promise to unify the fundamental interactions. However, the dazzling theoretical conceptions of string theory have persistently stumbled over significant hurdles pertaining to testability and falsifiability. This critical challenge casts a shadow on the theory's standing in the empirical realm of science.

Testability, a cornerstone of the scientific method, stipulates that for a theory to be scientifically valid, it must be capable of making predictions that can be tested and potentially falsified through experiments or observations. Falsifiability, a concept championed by philosopher Karl Popper, further demands that a theory must be structured in such a way that it can be refuted by evidence.

String theory posits that the elementary particles we observe are not point-like dots but rather tiny, vibrating strings. Each mode of vibration corresponds to a different particle type. The mathematical framework underlying this idea is remarkably elegant and extensive. However, the energy scales at which these strings ostensibly operate—near the Planck scale (about 10^{19} GeV)—are realms currently unreachable by our most powerful particle accelerators, such as the Large Hadron Collider (LHC), which operates at just about 13 TeV. This massive disparity in energy scales results in a scenario where direct experimental validation of string theory remains elusive.

Moreover, the 'falsifiability' aspect is compromised by the string theory landscape—a vast array of possible solutions to the string equations. This array comes from string theory's allowance of numerous different shapes and sizes for the extra dimensions it postulates. Estimates suggest there may be as many as 10^{500} different solutions. Each potential solution could describe a differ-

9.2. THE PROBLEM OF TESTABILITY AND FALSIFIABILITY

ent universe with different physical laws, thus complicating the predicament of making specific, falsifiable predictions about our universe.

Current experimental strategies indirectly approach the validation of string theory. One such avenue is the study of astronomical phenomena that might provide evidence of these extra dimensions or the unique signatures of stringy effects, like cosmic strings in the cosmic microwave background data. Another indirect test relates to the detection of supersymmetric particles, posited by certain string theory models, which could emerge within reach of future collider experiments.

Enhancing the dialogue between string theorists and experimental physicists is pivotal. Theorists are working on deriving "low-energy" implications from string theory that could manifest within accessible experimental ranges. Efforts are ongoing to refine or develop new non-accelerator-based experimental methods designed to explore higher energy scales more indirectly and economically.

Notwithstanding these efforts, the detachment of string theory's core principles from physical testing continues to attract substantial critique. Proponents argue that alternative methodologies in advancing the theory—such as increasing reliance on mathematical consistency and utilizing thought experiments (Gedankenexperiments)—remain valuable. Others posit that rather than traditional testability, string theory should perhaps be evaluated by its utility in generating novel mathematical frameworks and fostering deeper understanding in physics.

As we reflect upon these discussions, it becomes increasingly clear that the pathway to empirical relevancy for string theory is not through abandonment but through the arduous task of bridging theoretical foresight with innovative experimental ingenuity. The steadfast pursuit of such connections might one day either validate or compellingly challenge the precepts of string theory, thus upholding the rigorous standards of scientific inquiry.

9.3 Mathematical Complexity and Conceptual Understandability

When delving into the mathematical sophistication that string theory demands, one encounters a realm where elegance intersects with intricate complexity. The theory modifies our understanding of the universe at the most fundamental level, utilizing a mathematical framework that extends beyond the familiar three dimensions of space and one of time. To fully appreciate the depth and breadth of string theory, it is imperative to grasp its reliance on advanced mathematical concepts like Calabi-Yau manifolds, conformal field theory, and supersymmetry.

The backbone of string theory rests on the notion that point-like particles of particle physics are replaced by one-dimensional "strings" that vibrate at particular frequencies. Each vibration corresponds to a different particle, with its mass and force charge determined by the string's vibrational state. This apparently simple model quickly escalates in complexity as these strings operate in multi-dimensional spaces, requiring up to 10 or 11 dimensions to maintain mathematical consistency and facilitate the unification of all fundamental forces.

To comprehend these dimensions and their implications, consider the concept of Calabi-Yau manifolds. These are intricate, six-dimensional shapes essential for compactifying extra dimensions to a size so small as to be undetectable yet crucial for string theory's consistency. These manifolds are not arbitrary; they must satisfy specific mathematical properties to maintain supersymmetry, an aspect pivotal for canceling anomalies and inconsistencies within the theory.

Moreover, string theory incorporates the sophisticated framework of conformal field theory. This theory is pivotal as it describes the algebra of string interactions on a two-dimensional worldsheet – a surface swept out in time by a string – in a way that is independent of the specific geometry involved, focusing instead on how shapes can be stretched or shrunk while respecting angles. Conformal field theory is rich in symmetry properties, crucial for unraveling the dynamics of strings but demands a deep mathematical understanding to appreciate its full capacity and implications.

These theoretical constructs also challenge the conceptual understandability of string theory. Where classical mechanics and even general relativity were conceptualized through relatively comprehensible space-time diagrams or bending fabrics, string theory's abstract notions do not lend themselves as easily to such visualization. Moreover, the reliance on complex mathematical structures often makes the barrier to entry quite high for new students and researchers in the field. The general scientific community and the interested public face significant challenges grasping the foundational principles, making the dissemination and discussion of string theory somewhat constrained to those with advanced understanding of theoretical physics and higher mathematics.

In weaving through the dense forest of string theory's

mathematical complexity, scholars utilize cutting-edge mathematical techniques and increasingly sophisticated technology. The use of computer algorithms to handle complex calculations and visualize higher-dimensional theoretical constructs has become more prevalent, aiding in advancing understanding albeit incrementally. Despite these advancements, the immense complexity not only makes string theory a challenging field to master but also raises questions about the practicality of its broader comprehension.

Concept	Description	Challenge
Calabi-Yau manifolds	Six-dimensional spaces necessary for extra dimensions	High dimensional understanding required
Conformal Field Theory	Describes string interactions in 2D worldsheets	Deep symmetry and algebra knowledge needed
Supersymmetry	Theoretical framework, balancing fermionic and bosonic particles	Adds layers to dimensional and particle complexity

This tableau of theories and concepts makes it evident that the pathway to mastering string theory is paved with both awe-inspiring insights and formidable mathematical obstacles. It beckons those intrigued by the ultimate laws of the universe, promising a journey rich with both enlightenment and intellectual rigor.

9.4 The Landscape Problem: Multitude of Possible Solutions

String theory, with its beautiful mathematical scaffolding, proposes a tantalizing unified framework for understanding the universe's fundamental forces. However, one of its most debated aspects is the so-called "landscape" problem, which arises from the theory's prediction of a vast number (potentially around 10^{500}) of possible solutions or vacuum states. Each of these solutions corresponds to a different way in which the extra dimensions (presumed by string theory to exist) could be compactified or structured.

To better comprehend this issue, imagine each solution as a unique way to fold a much higher-dimensional space into a structure invisible to our everyday experiences yet crucial in determining physical constants and laws as we perceive them. The diversity in these compactification schemes leads directly to a diversity in possible universes, each with its own set of physical laws.

This multiplicity brings with it significant theoretical and empirical challenges. Theoretically, the existence of such a vast number of solutions prevents string theory from being predictive in the traditional sense used in physics. If a theory potentially predicts many different sets of physical laws, distinguishing the actual universe's laws becomes problematic without additional criteria or mechanisms to select among these numerous options.

Empirically, the multitude of plausible solutions complicates the possibility of verifying string theory through experimentation or observation. Normally, a scientific theory is deemed robust if it not only describes and predicts phenomena accurately but also does so uniquely or with manageable numbers of possibilities that can be systematically tested. In contrast, the sprawling landscape of string theory solutions makes it exceedingly difficult to test all feasible options comprehensively.

Mathematicians and physicists have sought ways to navigate this landscape via what is known as the "swampland" criteria. These criteria aim to establish conditions that a viable physical theory must meet to not end up in the "swampland" – a metaphor for theories that are consistent with mathematics but make no physical sense. The swampland criteria help theorists isolate more probable models from the string landscape's lesser, unphysical scenarios.

Despite these efforts, a central question remains tantalizingly open: How does one effectively select the correct vacuum state that corresponds to our universe from among countless candidates? Some theorists propose that principles like naturalness or anthropic considerations might guide this selection. The anthropic principle, for instance, suggests that our universe has the physical constants it does because if they were different, life and consequently observers capable of questioning these constants could not exist. This principle, however, is contentious and seen by many scientists as shifting away from predictive science towards post-hoc rationalization.

In graphical terms, if one were to represent each solution as a point in a high-dimensional space, what emerges is not a single clear dot but rather a vast "cloud" of possibilities stretching across conceptual realms. This visualization (not shown here but commonly represented in theoretical physics discussions) underscores the enormity of the challenge faced.

To progress from here, string theorists continue refining their mathematical tools and deepen their understanding of string theory's foundational principles, hoping to discover or invent mechanisms that naturally limit the landscape's expanse. Whether this endeavor will simplify or complicate the already intricate tapestry of string theory remains a fascinating aspect of modern theoretical physics, one that mirrors the broader debate about the nature of scientific inquiry and the pursuit of knowledge about our universe.

Thus, while the landscape problem presents significant theoretical hurdles and fuels much debate within the physics community, it also drives innovation in theoretical physics, prompting researchers to develop more sophisticated mathematical techniques and philosophical viewpoints in striving to understand the deepest workings of our cosmos.

9.5 Dependence on Higher Dimensions: A Double-Edged Sword

String theory posits the existence of additional spatial dimensions beyond the familiar three. According to the theory, these extra dimensions are compactified or curled up at scales so small that they remain unobservable with current experimental technology. This premise is intriguing as it opens avenues to unify all fundamental interactions of nature, including gravity, within a single theoretical framework. However, this reliance on higher dimensions also presents significant challenges and has become a double-edged sword in the development and acceptance of string theory.

The compactification of extra dimensions is a crucial aspect of string theory, as different modes of compactification can lead to different low-energy physical properties. Theoretically, this can result in a vast array of possible universes, each with its unique set of physical laws. This is not solely a speculative endeavor; rather, it is deeply rooted in the mathematics of string theory, particularly in the formalism that describes how strings vibrate and interact in these higher-dimensional spaces.

One of the primary implications of higher dimensions in string theory is their potential to explain phenomena that cannot be accounted for within the standard model of particle physics. For instance, the hierarchy problem, which questions why gravity is exponentially weaker than the other fundamental forces, can be addressed by models in which the gravitational force dilutes through larger extra dimensions. Moreover, the potential unification of forces through mechanisms like Kaluza-Klein theory, where the geometry of extra dimensions influences physical laws, presents a fascinating intersection of geometry and physics.

However, the mathematical beauty of these theories often clashes with practical scientific requirements, notably testability. The Planck scale, where these extra dimensions might manifest, lies far beyond current experimental reach. This predicament brings us to the crux of the issue: if we cannot test or observe these dimensions, can we truly consider their implications as scientific?

Further complicating this scenario is the issue of uniqueness. String theory's landscape suggests a multitude of possible solutions involving different shapes and sizes of extra dimensions. This multitude results in a lack of predictive power regarding the specific nature and effects of these dimensions. Notably, the absence of unique predictions makes it difficult to empirically validate the theory, thereby hindering its falsifiability—a cornerstone of traditional scientific methodology.

Critically examining these challenges, theorists have advanced several innovative approaches to mitigate these issues. Methods such as non-perturbative formulations, dualities, and the use of branes have been explored to provide more concrete predictions and potentially observable signatures that might indirectly confirm the presence of extra dimensions. For instance, certain models predict observable effects at the Large Hadron Collider (LHC), such as signatures of micro black holes or specific particle decay patterns indicative of extra-dimensional physics.

Beyond the realms of pure theory and toward potential experimental validation, efforts in astronomical observations and cosmic microwave background studies also offer indirect clues about the structure of these high-dimensional spaces. These efforts represent an intriguing bridge between high-level theoretical physics and observational cosmology, providing a tantalizing glimpse into how higher dimensions might manifest in observable phenomena.

As string theory progresses, it remains to be seen how these theoretical frameworks can be aligned more closely with empirical science. This alignment is critical not only for the advancement of string theory but also for the broader acceptance of higher dimensions as a fundamental part of our understanding of the universe. This conundrum epitomizes the perennial tension between the abstract beauty of theoretical physics and the empirical demands of the scientific method, a narrative that continues to unfold in the quest for a deeper understanding of the cosmos.

9.6 Lack of Unique Predictions: Challenges in Empirical Verification

One of the most critical aspects that sets scientific theories apart is their ability to predict phenomena which can then be empirically verified. String theory, despite its elegance and powerful mathematical framework, stumbles significantly in this arena. The theory posits a universe composed of tiny, vibrating strings whose modes of vibration manifest the particles observed in physical experiments. However, the exact predictions made by string theory can be varied and depend heavily on how the strings are theorized to exist and interact in multiple dimensions.

String theory's prediction dilemma primarily stems from its rich landscape of potential solutions. Each solution corresponds to a different set of physical laws, effectively implying that virtually any observation in particle physics could potentially be explained by one of these solutions. This multitude of solutions is not merely vast but is theoretically so extensive that pinning down one specific prediction that can be tested universally becomes challenging.

To illustrate this point, let us consider the energy scale at which string theory operates. String theory's entities, like strings and branes, exist at scales close to the Planck length, which is approximately 10^{-35} meters. This is a realm so tiny and energy-intensive that current technological apparatus cannot probe such energetic interactions. The Large Hadron Collider (LHC), our most powerful particle accelerator, operates at energies around 10^{12} to

9.6. LACK OF UNIQUE PREDICTIONS: CHALLENGES IN EMPIRICAL VERIFICATION

10^{13} electronvolts, which pale significantly in comparison to the energies of about 10^{19} GeV required to test string-theoretic predictions directly.

Implicit in these challenges is the issue of mathematical framing. String theory, enriched with complex mathematical structures such as algebras, differential geometry, and topology, often results in equations that do not limit their solution space effectively enough to provide distinct predictions. Often, precise calculations in string theory lead to a multitude of possible outcomes, influenced by differing compactifications of the extra dimensions string theory posits. For example, how the six extra dimensions proposed by String Theory are compacted can lead to various "low-energy effective" theories, each compatible with observed physical phenomena but distinctively different at untested higher energies.

Moreover, the dependence on supersymmetry adds another layer of complexity. Supersymmetry, a principle that predicts a partner particle for every particle existing in our universe, helps to stabilize many of the mathematical inconsistencies of string theory. However, despite years of searching, experiments such as those conducted at the LHC have yet to confirm the existence of these supersymmetric partners, resulting in further skepticism regarding the empirically verifiable predictions of string theory.

Despite these challenges, it is important to reflect on the nature of theoretical progression in physics. Historically, theories often undergo numerous modifications and face periods where empirical verification remains elusive. For instance, the bending of light prediction by General Relativity waited years for eclipse observations for its verification. Similarly, while string theory currently faces significant empirical hurdles, this does not fatally undermine its utility as a framework in theoretical physics.

Engaging with string theory's predictions thus requires not only sophisticated technological advancement but also a sustained commitment to theoretical exploration under conditions of profound uncertainty. As such, experimental tests for string theory are less about direct confrontation and more about indirect consistency checks with the plethora of data emerging from other fields like astrophysics and cosmology. These observations could potentially limit or specify the "correct" version of string theory applied to our observable universe.

Hence, grappling with the veneration of empirical verification in the age of theoretical complexity like that posed by string theory invites a broader question about the thresholds of understanding and technological prowess at which invisible realms of cosmic strings become part of observable reality. The detailed interplay between prediction and verification in string theory not only challenges our experimental ambitions but also enriches our philosophical discourse on the nature of scientific truth and progress.

9.7 Competing Theories: Loop Quantum Gravity and Others

As string theory continues to grapple with various criticisms regarding its testability, other competing theories also bid to solve the profound puzzles of quantum gravity and cosmology. Among these, Loop Quantum Gravity (LQG) stands out as a noticeable contender, introducing distinct approaches and ideas to the discourse of theoret-

9.7. COMPETING THEORIES: LOOP QUANTUM GRAVITY AND OTHERS

ical physics. This section delves into the key aspects of LQG, juxtaposes it against string theory, and briefly discusses other theories like Twistor Theory and Causal Dynamical Triangulation, broadening our horizon beyond the realms string theory inhabits.

Loop Quantum Gravity primarily differs from string theory in that it does not introduce additional dimensions or rely on the concept of supersymmetry. LQG attempts to quantize spacetime itself, unlike string theory, which replaces the point-like particles with one-dimensional strings to unify gravity with quantum mechanics at the smallest scales. The foundational aspect of LQG is that space is not continuous but composed of tiny loops woven into a fine fabric that represents the gravitational field. These loops form a network or graph, known as a spin network, which evolves over time into what is known as a spin foam—the quantum equivalent of a spacetime.

Aspect String Theory	Loop Quantum Gravity
Basic Framework Uses 1-dimensional strings to unify forces, often requiring extra dimensions.	Quantizes spacetime itself using loops. No need for additional dimensions.
Mathematical Tools Often relies on conformal field theory and algebraic geometry in higher dimensions.	Uses background-independent quantum theory methods.
Theoretical Outputs Predicts the existence of multiple, possible universes and often ties into the idea of a multiverse.	Predicts discrete space-time structure. Creates possible insights into the early universe's quantum states.

Moreover, another potential competitor to string theory is Twistor Theory, which emerged from the work of Sir Roger Penrose. Unlike string theory and LQG, Twistor Theory reformulates the field equations of general relativity using complex variables in a twistor space – an approach that potentially offers new perspectives on space-time and interactions. This theory also aims at unification but does so through an entirely different mathematical landscape.

Causal Dynamical Triangulation (CDT) is yet another approach, seeking to provide a background-independent theory of quantum gravity. This theory notably stands out by imposing a piecewise linear spacetime that captures dynamic topological changes without relying on fixed coordinates—marking another stark departure from both string theory and LQG.

Understanding these nuanced differences is essential as it illuminates the diverse ways physicists approach the enigma of quantum gravity, illustrating that the theoretical physics community thrives on a pluralistic approach to solving its most intricate problems. Our journey through these theories not only shows alternative solutions but also refines our philosophical understanding of what a theory of everything could possibly entail—acknowledging that our exploration of cosmic truths remains wonderfully incomplete.

9.8 Philosophical Implications: Science or Philosophy?

The dichotomy between science and philosophy in the context of string theory is a conundrum that potently encapsulates the broader philosophical debates within modern physics. At the heart of this discussion is the fundamental question: is string theory, with its current lack of direct empirical evidence and high-dimensional mathematical framework, more a philosophical construct than a scientific endeavor?

String theory offers a framework to unify all fundamental

CHAPTER 9. CHALLENGES AND CRITIQUES OF STRING THEORY

forces and forms of matter in a single theoretical structure, but it does so in a manner that stretches the conventional criteria of empirical validation. Karl Popper's philosophy of science argues that for a theory to be scientific, it must be falsifiable. In other words, there must exist a possible experiment whose negative outcome would definitively reject the theory. However, string theory, as currently posited, operates at energy scales so high that today's technology is inadequate to test its predictions directly. This challenge leads some critics to question whether it falls within the domain of scientific inquiry or wanders into metaphysical speculation.

To address these concerns, some philosophers and scientists propose that we consider a new criterion of 'theoretical confirmation' beyond direct empirical falsifiability. This perspective relies heavily on the theory's internal consistency and its ability to unify disparate phenomena under a common framework. For example, string theory's elegance in explaining both quantum mechanics and general relativity with a single apparatus could be viewed as a form of indirect empirical support. The coherence between string theory and known physics adds a layer of philosophical complexity, forcing a reevaluation of what we consider empirically verifiable science.

Moreover, string theory brings to the forefront the issue of mathematical platonism—the belief that mathematical entities exist independently of physical events. This philosophy is evident when theorists explore the vast landscape of string theory solutions, each describing different possible universes with varying laws of physics. Are these solutions mere mathematical artifacts, or do they exist in a real, albeit non-empirical sense? This intertwining of mathematical existence and physical reality challenges the traditional boundaries of science and philosophy, urg-

9.8. PHILOSOPHICAL IMPLICATIONS: SCIENCE OR PHILOSOPHY?

ing a synthesis of both realms.

Additionally, the multiverse concept, often linked with string theory, raises further philosophical questions. If our universe is just one of many in a giant cosmic landscape, can science hope to empirically assess realities beyond our observational capacity? This leads to a provocative debate over whether such theories can ever be scientific if they do not yield testable predictions about our specific universe.

In the bigger picture, string theory's critiques and the surrounding debate shine a light on the evolving nature of scientific theories. Historically, shifts in scientific thought often blur the lines between established empirical science and philosophical speculation. As such, rather than diminishing the value of string theory, these philosophical debates could be seen as indicative of deep, meaningful progress—pushing the boundaries of how we understand and define scientific truth.

The discourse surrounding string theory and its philosophical implications suggests that perhaps it is time to broaden our conception of what science can be. Rather than adhering strictly to traditional empirical methodologies, we might consider embracing a more pluralistic approach to scientific validation, one that accommodates the uniquely speculative yet profoundly unifying aspirations of theories like string theory.

9.9 Evolving Theories: Adaptability and Modification of String Theory

String theory, in its essence, is not merely a set of static principles, but it is a dynamic and evolving framework, continually refined and adapted in response to new mathematical insights and experimental findings. The adaptations, whether they are refinements in the mathematical formalism or theoretical extensions, serve as the vibrant underpinnings that fund its resilience and capacity for progression amidst scrutiny.

Initially posited as a framework for understanding the strong nuclear force, string theory underwent its first significant evolution with the advent of supersymmetric string theory, or superstring theory, in the 1970s. This incorporation of supersymmetry was instrumental in broadening the implications of string theory from not just a theory of nuclear force but as a conceivable theory of everything, potentially explaining all particles and fundamental forces.

Adapting Mathematical Tools: Central to the adaptation of string theory is the expansion of mathematical instruments used to tackle the enormously complex issues it contemplates. Tools from algebraic geometry, topology, and differential geometry have been particularly influential. For example, the Calabi-Yau manifold became significant in the context of compactifying additional dimensions posited by string theory into a scale that could align with our observable universe.

The use of dualities, another crucial adaptation in string theory, provides a powerful theoretical instrument that allows physicists to relate different string theories which at first seem distinct. Dualities have led to great simplifications in understanding conditions under strong coupling (where forces are strong and perturbation theory fails) by linking them to equivalent weak coupling conditions, where calculations are more feasible. This concept manifested through various forms of duality such as T-duality and S-duality, expanded the robustness of string theory to include a web of interlinked theories, suggesting a possible underlying unification.

Addressing the Non-Perturbative Regime: Adaptations to string theory have also expanded its applicability beyond perturbative areas. The development of brane dynamics within the framework addressed points where the earlier versions of string theory could not effectively tread — specifically in the non-perturbative regime where the interactions are notably strong. Branes — multidimensional objects embedded within higher-dimensional space — opened avenues to explore phenomena like black hole physics within the mantle of string theory, linking it with quantum gravity.

Influence on Quantum Gravity and Holography: Furthermore, these developments contributed significantly to visions of a holographic universe — an approach collaborated upon through the anti-de Sitter/Conformal Field Theory (AdS/CFT) correspondence. This principle implies that a quantum gravity theory defined on a particular space can be equivalent to a conformal field theory on the boundary of that space, providing a powerful toolkit for studying quantum gravity theoretically.

Response to Empirical Challenges: As experimental results continue to probe deeper into the foundational layers of physical reality — such as observations from particle accelerators and cosmological data — string theory continues to adapt. For instance, adjustments to string compactifications, which define how extra dimensions might be curled up, are continually modified in light of observational cosmology, pushing the theory towards possible testable predictions.

These modifications show that string theory is a rich framework capable of powerful adaptation and theoretical resilience. While the path forward is laden with empirical challenges and conceptual hurdles — such as proving the uniqueness of its solutions or directly linking its predictions with experimental data — the progress in its adaptability through mathematical and conceptual evolution suggests a future where it may yet provide definable connections to observed phenomena. Its journey exemplifies the vibrant narrative of theoretical physics — where adaptation is not just a response but an advance towards deeper understanding.

9.10 Future Paths: Addressing the Critiques Effectively

String theory, a frontrunner in the race for a Theory of Everything, has indeed confronted significant hurdles, not least of which involves its testability, the seemingly endless multitude of potential solutions, and its reliance upon unobservable additional dimensions. Yet, the forward momentum in theoretical physics demands innovative and aggressive approaches to either validate or reformulate the theory with higher degrees of empirical engagement. Addressing the critiques of string theory effectively revolves around several pivotal strategies aimed at

9.10. FUTURE PATHS: ADDRESSING THE CRITIQUES EFFECTIVELY

enhancing its scientific rigour and relevance.

First, ongoing and intensified collaboration between theoretical physicists and experimentalists is crucial. Concrete strides have been made in designing experiments that can probe energy scales much beyond those available today, despite none being able to directly test string theory yet. Projects like the Large Hadron Collider (LHC) have nudged us closer to conditions prevalent just after the Big Bang, yet more sensitive equipment is needed. Future colliders or new technologies such as quantum computers might hold the key to simulating and hence testing predictions derived from string theory.

Advancements in observational astrophysics have also shown promising auxiliary routes. The study of gravitational waves, for instance, offers an indirect arena where the predictions of string theory might manifest. These ripples in spacetime fabric, generated by cataclysmic astronomical events such as black hole mergers, might carry subtle signatures predicted by string theory adaptations. Enhancing the sensitivity of instruments like LIGO and collaborating globally on projects like the Event Horizon Telescope could amplify our capabilities to capture potential evidence supporting or refuting aspects of string theory.

Enhancement of the theoretical framework itself to reduce the landscape problem is another critical area. Efforts to streamline the multiverse of solutions in string theory involve refining criteria or developing new mathematical tools to prune the vast landscape. Methods such as the Swampland criteria aim to delineate effective field theories that can arise from a string theory consistently with quantum gravity. This approach helps in discerning which low-energy effective theories are compatible with string theory and which are not, potentially reducing the plethora of possible string theory solutions to a manageable few that can be systematically explored further.

Moreover, embracing alternative theories and integrating insights from them can provide fresh perspectives that might either complement or challenge string theory's postulates. Engaging with other leading theories like loop quantum gravity or causal dynamical triangulations opens up interdisciplinary pathways where components from each model may synergize to address the fundamental questions of physics. Working beyond the confines of string theory exclusive communities fosters a broader theoretical landscape and stimulates innovative approaches.

Additionally, public communication and philosophy of science play a supportive role in addressing critiques related to the scientific legitimacy of string theory. Enhancing the discourse around what constitutes acceptable scientific evidence, revising paradigms on the demarcation between science and non-science, and effectively communicating these shifts to the broader public are essential. This approach will help in managing expectations and clarifying misunderstanding surrounding the status and objectives of string theory within the scientific community and beyond.

Through an amalgamation of heightened empirical approaches, refined theoretical advancements, interdisciplinary collaboration and philosophical introspection, string theory can confront its critiques head-on. While these pathways illuminate a challenging road ahead, they embody the robust dynamism inherent in scientific endeavor — forever adapting, evolving, and striving towards a deeper understanding of the universe.

Chapter 10

The Future of String Theory: Theoretical and Experimental Prospects

This chapter explores the prospective advancements and ongoing research in string theory, both from theoretical and experimental viewpoints. It discusses anticipated developments in the mathematical tools and concepts integral to string theory, as well as emerging technologies that may enable new experimental tests. The narrative also considers how collaborations across various domains of physics, such as quantum computing and cosmology, could influence the future trajectory of string theory research. Additionally, it assesses the potential for future particle accelerators and observational facilities to provide data that could further elucidate or challenge the premises of string theory.

10.1 Current State and Immediate Future Directions in String Theory

Delving into the present landscape of string theory, we recognize its standing as a pivotal theoretical framework in high energy physics and cosmology. String theory has matured over decades, offering a profound unification of quantum mechanics and general relativity. At its heart, string theory posits that point-like particles are replaced by one-dimensional strings, whose modes of vibration manifest as various elementary particles.

Current research in string theory remains richly theoretical, focusing predominantly on refining mathematical formulations and overcoming conceptual hurdles. One central challenge is the multitude of solutions allowed by string theories—a landscape potentially consisting of 10^{500} different universes, each with its own set of physical laws. This "landscape problem" underscores the need for a principle or mechanism to select the universe that corresponds to our own.

Moreover, another immediate focus lies in resolving ambiguities associated with dualities and gauge symmetries. Dualities in string theory suggest that seemingly distinct physical theories could be mathematically equivalent under certain conditions, challenging our understanding of fundamental principles.

Looking ahead, the immediate future involves multiple theoretical fronts:

- **Enhanced Mathematical Tools:** New computational methods and enhanced algebraic structures are pivotal. This would include advanced techniques in algebraic geometry and number theory aimed at solving complex differential equations intrinsic to string theory.

- **Formalizing String Field Theory:** Efforts to formalize string field theory continue with the aim to provide a coherent quantum field theory of strings. This advancement could bridge the lesser-understood dynamics of string interactions.

- **Holographic Principle Applications:** Inspired by the anti-de Sitter/conformal field theory (AdS/CFT) correspondence, applications of the holographic principle are becoming crucial. This principle provides powerful insights into quantum gravity and even areas outside traditional physics, like information theory and quantum computing.

- **Quantum Gravity Insights:** A deeper understanding of quantum aspects of gravity, fostered by advancements in string theory, remains a priority. The promise of obtaining a consistent quantum theory of gravity is one of string theory's most exciting prospects.

Integrated into these theoretical advancements is the parallel progression thanks to technological advancements in computational fields. Supercomputing power significantly boosts the ability to handle the extensive numerical simulations required for testing string theory implications and exploring new theoretical landscapes. Additionally, advancements in machine learning might provide new ways to explore the string landscape, identifying patterns or predictions that are not readily apparent through conventional approaches.

The immediate future also holds potential for exciting collaborations across different sectors of theoretical physics and practical experiments. One such intersection is with the phenomenology of particle physics, where developments might soon provide low-energy footprints of string theory through observable signatures like magnetic monopoles or cosmic strings.

The creative synergy between abstract theoretical pursuits and empirical investigations summons a vibrant future for string theory, demonstrating its foundational role in understanding the intricacies of the cosmos. As this exploration forges ahead, the continuity from current state to near future directions in string theory becomes a fascinating narrative of discovery and intellectual triumph within the scientific community.

10.2 Advancements in Mathematical Formulations and Techniques

The endeavor to delve deeper into the fabric of the universe through string theory necessitates continual refinement and development of mathematical tools and concepts. This drive stems from the unique demands of string theory, which intertwines intricate geometry, topology, and algebra. The leaps in understanding this complex theory significantly depend on emerging mathematical formulations that can accurately describe and predict the properties of strings and their interactions in multiple dimensions.

One pivotal area of advancement is in the realm of algebraic geometry and its applications to string theory. Complex manifolds used in string theory, particularly Calabi-

10.2. ADVANCEMENTS IN MATHEMATICAL FORMULATIONS AND TECHNIQUES

Yau manifolds, have been a cornerstone for many theoretical insights. Recent developments have focused on better understanding the moduli spaces of these manifolds, which represent the different ways a manifold can be deformed while still retaining its essential properties. Enhanced computational tools have now enabled theorists to catalog properties of a vast array of Calabi-Yau manifolds, potentially leading to a more profound understanding of string landscape and vacuum stability.

The application of homological mirror symmetry provides another illustration of advancing mathematical techniques. This concept acts as a bridge between algebraic and symplectic geometry; it allows the translation of problems in enumerative geometry into equivalent problems in symplectic topology. Recent efforts have been directed toward proving cases of the homological mirror symmetry conjecture, with significant implications for both mathematics and string theory.

Further, integration of techniques from quantum algebra such as vertex algebras has substantially enriched our understanding. Vertex algebras, crucial in the conformal field theory aspect of string theory, facilitate detailed analysis of boundary conditions and D-branes in string models. Developments in this area are potent enough to enhance our comprehension of string dynamics in lower dimensions and provide a pathway to systematically address higher-dimensional questions.

Significant strides have also been made in the use of machine learning and artificial intelligence to tackle complex problems in string theory. These technologies have started to play a role in areas like the classification of string vacua and in the calculation of scattering amplitudes, realms that involve high dimensional data analysis and pattern recognition. The integration of these computational technologies opens up new vistas for handling the substantial complexity and volume of calculations required in string theory.

Moreover, topological field theories continue to offer profound insights into the non-perturbative aspects of string theory. The refinement of topological vertex techniques, which are essential for the computation of Gromov-Witten invariants, underscores how upgrades in mathematical methods can have direct implications on physical predictions in string theory. The ongoing development of field-theoretic methods to better understand the link between string theory and quantum gravity is another frontier being actively explored.

Each of these developments fortifies the mathematical backbone of string theory, enabling a more nuanced exploration into the theory's predictions and its correspondence with physical reality. By marrying traditional mathematical methods with innovative computational techniques, the field is poised to unlock new realms within both mathematics and physics, continually pushing the boundaries of what we comprehend about the underlying principles of the universe.

10.3 Emerging Technologies and Their Impact on Experimental String Theory

The pioneering surface in string theory, historically a predominantly theoretical framework due to its intricate high-energy implications, is witnessing profound trans-

10.3. EMERGING TECHNOLOGIES AND THEIR IMPACT ON EXPERIMENTAL STRING THEORY

formations through emerging technological advancements. Today, cutting-edge technologies not only refine theoretical predictions but also bridge them with tangible experimental investigations—ushering in a new era where theoretical constructs once considered far beyond empirical reach are increasingly testable.

A key advancement is in the realm of precision measurement tools. Enhanced capabilities in laser interferometry, for instance, are crucial for detecting minuscule perturbations in spacetime induced by gravitational waves. These tools could potentially be sensitive enough to detect oscillations or anomalies predicted by certain models of string theory. Advanced interferometers, utilizing quantum squeezing to reduce noise below the standard quantum limit, mark a significant leap forward in this domain. The ability to measure at such precision opens new frontiers for testing the principles of string theory, particularly in scenarios involving microscopic black holes and cosmological string vibrations.

Further afield, the development of more powerful particle accelerators represents another cornerstone for probing string theory. While the Large Hadron Collider (LHC) has provided invaluable insights into high energy particle physics, future accelerators such as the proposed Future Circular Collider (FCC) or the International Linear Collider (ILC) are expected to reach even higher energies. These facilities aim to explore energy scales where conjectured phenomena like supersymmetry and extra dimensions—central to string theory—are possible. By smashing particles at unprecedented energies, these accelerators could uncover evidence of stringy effects at smaller scales than currently observable.

The integration of quantum computing in experimental physics is another transformative technology. Quantum computers offer immense potential to simulate quantum phenomena that are profoundly complex for classical computers, such as the superposition and entanglement principles intrinsic to quantum field theories and string theory. Tools developed from quantum computing could allow physicists to simulate and study string theory scenarios in controlled, quantifiable settings, providing a new approach to understand string theory's implications on quantum gravity and the unification of forces.

Additionally, developments in observational astronomy and astrophysics provide unexpected avenues to test string theory. The study of high-energy cosmic rays, for instance, offers data on particle interactions at energies above those achievable in man-made accelerators. As string theory often provides frameworks that extend beyond standard models of particle physics, observations of cosmic phenomena can offer indirect yet valuable insights into extra-dimensional spaces and other string-related phenomena.

From a technology standpoint, innovations such as AI and machine learning are being enlisted to manage the sheer volume of data generated from experiments relevant to string theory predictions. Intelligent algorithms can discern patterns and anomalies which might indicate new physical phenomena corresponding with string theory's predictions, vastly improving the efficiency and scope of potential discoveries.

In light of these technological revolutions, experimental string theory is transitioning from a nascent stage to one of active exploration and testing. As we forge ahead, the synergy between advanced technology and pioneering theoretical physics continues to redefine the limits of what can be discovered about the very fabric of our universe. Each breakthrough not only enhances our understanding of string theory but also reinforces the intricate connection between multiple fields of science, from particle physics to cosmology, creating a compound framework where theoretical constructs and practical experimentation coalesce into a cohesive understanding of universal laws.

10.4 Integration with Quantum Computing and Information

The synchronization of string theory with quantum computing and information technology is an intriguing avenue that might significantly enhance our understanding and handling of data in theoretical physics. With quantum computing's unprecedented ability to manipulate and process information through quantum bits or qubits, the issue of complex calculations in string theory becomes a more approachable challenge.

String theory, with its requirement for 10 or 11 spacetime dimensions and its complex landscape of solutions, poses significant computational challenges. Conventional computing resources are often strained by the sheer number of variables and the quantum nature of the computations involved. Here, quantum computers enter the fray, providing a novel approach to these problems through their inherent capability to perform operations on quantum data states simultaneously due to superposition, and to link operations through quantum entanglement.

The pivotal aspect of this integration lies in the encoding of string theory's data and equations into a quantum architecture. Using qubits instead of classical bits allows for a possible exponential increase in processing power, ideally suited for exploring string theory's vast landscape. For example, the AdS/CFT correspondence, a popular conjecture in string theory that associates a gravity theory in a higher-dimensional Anti-de Sitter space with a conformal field theory in one lower dimension, could substantially benefit from quantum simulations. These simulations can potentially model gravitational phenomena or explore quantum gravity scenarios within a controlled, error-tolerant quantum computing framework.

Moreover, the development of quantum algorithms tailored for string theoretical models can lead to breakthroughs not just in theoretical physics but also in optimizing quantum computing methodologies itself. This bi-directional benefit implies that while we use quantum computers to decipher string theory, the theoretical challenges posed by string theory can concurrently inspire novel quantum computational methods.

Recent advances in topological quantum computing offer a concrete example of beneficial overlap between fields. Topological quantum computers, which compute using quasiparticles called anyons, operate in ways that mimic the theoretical constructs of string theory, especially in higher-dimensional spaces. These systems are inherently error-resistant due to their topological nature, making them ideal candidates for tackling the complex calculations required by string theory without substantial loss of data integrity.

Practical implementation of such integration, however, faces notable hurdles. The creation and maintenance of qubits in a state coherent enough to perform calculations, error correction in quantum calculations, and the development of scalable quantum computing infrastructure are critical challenges that need addressing. Additionally, the theoretical aspect demands a deep understanding of both quantum mechanics and string theory, bridging knowledge gaps that traditionally exist between physicists specializing in these disparate fields.

Despite these challenges, the integration of string theory with quantum computing holds promising potential. It not only offers a new paradigm for testing and understanding string theory but also pushes the envelope in quantum computing, providing a unique insight into how these technologies might evolve symbiotically.

As quantum technologies and string theory both advance, their intersection could reveal deeper secrets of the universe, possibly leading to a new era in both fields where questions about the fundamental nature of reality are addressed not merely theoretically but through empirical quantum-supported frameworks.

10.5 Colliding Worlds: String Theory Meets Phenomenology

String theory, since its inception, has largely been a theoretical framework, one that is highly mathematical and somewhat removed from direct experimental validation. Yet, the burgeoning field of phenomenology in string theory presents an intriguing frontier where the inchoate predictions of string theory begin to engage with the empirical universe. This convergence of string theory with phenomenology bridges the abstract realms of high-dimensional theories with observable phenomena, offering tantalizing possibilities for verifying or refuting various aspects of string theory.

Phenomenologists have begun to adapt and apply the complex framework of string theory to solve specific problems in particle physics and cosmology. This adaptation often involves deriving viable models from string theory that could potentially be tested using real-world data. A successful model needs to not only address known issues like the hierarchy problem or the nature of dark matter but should also be compatible with existing experimental results and capable of predicting new, verifiable phenomena.

One approach in this interdisciplinary endeavor is the construction of string theory landscapes. These landscapes describe a vast multitude of possible vacuum states which string theory could inhabit. Researchers have employed statistical methods to analyze these landscapes, seeking regions that resemble our own universe. Each point in this landscape potentially corresponds to a different set of physical laws or constants. By comparing these predictions with known physical laws and constants, phenomenologists aim to find a configuration that matches our observable universe.

For example, the quintessential cases of flux compactifications have shown promise in explaining the stabilization of extra dimensions, a feature predicted by string theory yet unobserved directly. These compactifications involve intricate calculations where six of the ten required dimensions in superstring theory are 'compactified' on geometrically complex spaces known as Calabi-Yau manifolds. The properties of these manifold shapes can lead to different physical properties, like the values for the masses of particles and the strengths of forces, which are tuned to match actual observations in particle physics.

10.5. COLLIDING WORLDS: STRING THEORY MEETS PHENOMENOLOGY

Linking string theoretical predictions with particle collisions in large-scale facilities like the Large Hadron Collider (LHC) provides an experimental testbed for string phenomenology. Through high-energy collision data, phenomenologists have investigated signatures of supersymmetry, a principle deeply embedded in many string theory models. While direct evidence of supersymmetry has yet to be found, refined theoretical models and improved experimental technologies continue to refine our search parameters.

On the cosmological front, the implications for early universe models are profound where concepts such as string inflation provide a unique telescope to peer back at the inaugural moments following the Big Bang. String thery's version of inflation involves higher-dimensional dynamics, distinct from traditional inflationary models, potentially leading to observable effects such as distinctive patterns in the cosmic microwave background radiation.

Moreover, the quest to unify gravitation with other fundamental forces courtesy of string theory has iteratively borrowed concepts from phenomenology. This synergy has led to enhanced theoretical frameworks where phenomena such as black hole entropy and Hawking radiation are approached through a string theory lens, with each prediction opening new experimental or observational fronts.

As we continue to refine these ideas and challenge them against empirical data, we are drawing the map that connects these once distant worlds of abstract mathematics and tangible physics. The potential to one day corroborate portions of string theory experimentally continues to drive innovative theoretical developments and inspire a more holistic view of the cosmos. By informally auctioning off pieces of the theoretical structure for empirical testing, we may eventually articulate a more coherent narrative that aligns closely both with the mathematics at the heart of string theory and the universe it seeks to describe.

10.6 Prospects for Uncovering Extra Dimensions

Among the most exciting concepts in string theory is the hypothesis of extra spatial dimensions beyond the familiar three. The implications of successfully identifying extra dimensions reach far beyond the remit of theoretical aesthetics—they offer tangible insight into the fundamental structure of the universe and bridge numerous gaps in current understanding within high energy physics, particularly in unifying general relativity with quantum mechanics.

Extra dimensions are hidden from our everyday experience because they are compactified, or curled up, at scales possibly as tiny as the Planck length, which is around 1.62×10^{-35} meters. Detecting them, therefore, demands precision and innovation in experimental physics and a rethinking of existing theoretical frameworks. Current strategies focus on indirect impacts observable through regular dimensions, such as deviations in gravitational behavior or consequences in particle physics phenomenology.

Gravity Tests and Collider Experiments One prominent approach to probing extra dimensions is through tests of gravity at miniature scales. Techniques have

10.6. PROSPECTS FOR UNCOVERING EXTRA DIMENSIONS

been designed to detect changes in gravitational force at distances shorter than a millimeter—an anomaly at this scale might suggest gravitational leakage into extra dimensions. Future experiments are planned to push this boundary even closer to the Planck scale. Additionally, particle colliders such as the Large Hadron Collider (LHC) offer possibilities to observe phenomena implying extra dimensions. For instance, the production of micro black holes during high-energy collisions would be indicative of gravity's behavior in more than three spatial dimensions.

To clarify the relationship between energy scales and potential discoveries at colliders, consider the following diagram depicting collider energy capabilities relative to predicted phenomena:

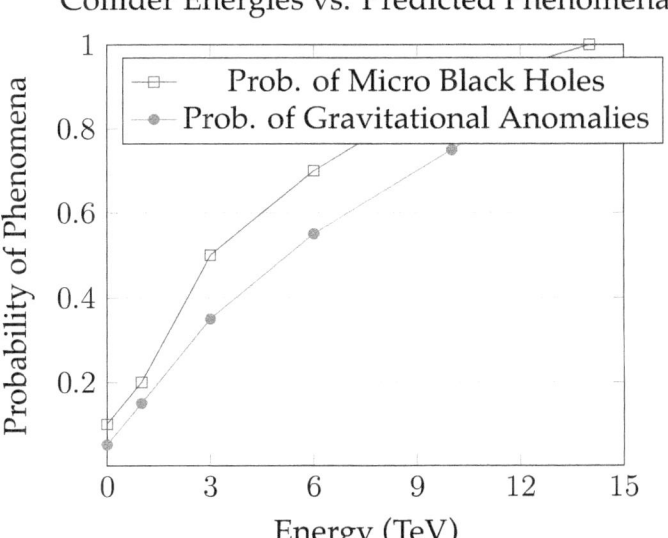

Influence of Quantum Computing Recent advances in quantum computing also provide a new frontier for testing ideas related to extra dimensions. Theoretically, a sufficiently powerful quantum computer could simulate high-energy physics scenarios at energies unattainable in current collider experiments. This is particularly relevant for studying the impacts of string theory, where the ability to manipulate and observe quantum states directly can simulate conditions closer to those at the inception of the universe.

Astrophysical Observations Additionally, precision measurements in cosmology might offer indirect clues about the geometry of the universe that suggest extra dimensions. Studies monitoring cosmic microwave background (CMB) anisotropies and large scale structure formations provide constraints on theories postulating extra dimensions.

As the frontier of technology advances alongside theoretical insights, the prospective detection and characterization of extra dimensions grows increasingly feasible. The confluence of interdisciplinary approaches and international collaborations will significantly enhance our capability to explore these profound questions about the nature of reality, potentially heralding new paradigms in physics and beyond.

10.7 New Generations of Particle Accelerators and Detectors

The exploration of string theory at the experimental level has historically been thwarted by the theory's predication on energy scales far beyond those accessible to current experimental technology. However, emerging designs and proposed implementations for new generations of particle accelerators and detectors hold promise for bridging this gap, potentially unlocking insights into the rarefied realms where string theory operates. These advancements involve both upgrades to existing facilities and plans for revolutionary new tools capable of probing energies that were previously unattainable.

Let us begin with the innovations in collider design. The Future Circular Collider (FCC), a concept proposed by CERN, aims to succeed the Large Hadron Collider (LHC) by offering an unprecedented collision energy of up to 100 TeV, which is nearly ten times more potent than the LHC. This leap in energy could be vital for testing predictions of string theory such as the existence of supersymmetric particles or extra spatial dimensions. The theoretical potential of the FCC in exploring these high-energy realms provides not just opportunities for verification of string theory but also a broader scope in the search for new physics.

Equally important in the new generation of accelerators is the International Linear Collider (ILC), which contrasts with the circular design of the FCC and LHC. The ILC would enable highly precise measurements that are less achievable in circular colliders due to their inherent emission of radiation by charged particles in curved trajectories. This linearity minimizes energy losses, allowing the ILC to focus on upscaling the precision of tests rather than energy, offering another vital perspective in probing theoretical predictions by string theory.

In addition to energy scale innovations, advancements in detection technologies are critical. New particle detectors are featuring increased granularity and sensitivity. Such improvements are essential for detecting elusive signals that might indicate phenomena promised by string theory, such as minute discrepancies in standard model predictions or the indirect effects of virtual particles hypothesized by string theory.

A significant factor in these advancements is the integration of emerging technologies such as machine learning and advanced computational algorithms. These tools are revolutionizing the way data from particle collisions is analyzed. For instance, machine learning algorithms can sift through petabytes of data to isolate potentially interesting events from the background noise, a task that is increasingly challenging at higher energies and intensity of particle collisions.

The role of quantum technologies in enhancing the capabilities of detectors should not be understated. Quantum sensors, for instance, utilize principles of quantum mechanics to achieve sensitivities in measuring physical quantities at levels previously deemed impractical. This heightened sensitivity could play a crucial role in detecting faint signals from new high-energy processes influenced by string theory dynamics.

As the development of these new generations of particle accelerators and detectors moves forward, it becomes evident that their contribution to the field of string theory and fundamental physics will be profound. Through these remarkable tools, the veiled sectors of the universe hinted at by string theory may soon yield their secrets, offering a clearer view into the very fabric of the cosmos.

These ongoing advancements encapsulate not just technological triumphs but also a reinvigorated hope for answers to some of the most profound questions in theoretical physics. Through the lenses of these powerful new machines, our understanding of the universe's fundamental nature is poised for exciting revelations. The interplay between theoretical predictions and experimental explorations continues to weave an ever more intricate tapestry, emblematic of the vibrant and dynamic field of string theory.

10.8 Interdisciplinary Approaches: From Cosmology to Condensed Matter

Consider the rich tapestry of the cosmos, a universe filled with mysteries that are just begging to be unraveled. String theory, originally conceived as a theory of the tiniest constituents of matter, has burgeoned into a framework with profound implications across various branches of physics. One of the most intriguing aspects of contemporary string theory research is its potential for interdisciplinary application, notably between the seemingly disparate realms of cosmology and condensed matter physics.

The relationship between cosmology and string theory is well-established, given that string theory attempts to integrate gravity with the other fundamental forces, therefore offering a coherent description of the early universe and its evolution. Recent advancements in this area leverage the symmetries and dualities inherent in string theory to tackle cosmological questions such as the nature of dark energy and black hole singularities. For instance, the use of Anti-de Sitter/conformal field theory (AdS/CFT) correspondence has allowed physicists to explore holographic descriptions of the universe that reconcile quantum mechanics with general relativity.

Turning our focus to condensed matter, we enter a realm traditionally concerned with the properties and behaviors of bulk systems and materials — a field that seems, at first glance, far removed from the lofty realms of cosmological phenomena. However, the emergent concept of 'holographic duality' from string theory has found unexpected success in describing phenomena in strongly correlated quantum systems, such as high-temperature superconductors and quantum critical points. In these systems, the AdS/CFT correspondence provides a powerful toolkit for modeling features that are typically intractable with conventional quantum field theory methods.

The linkage goes deeper. Theoretical tools developed in string theory are being applied to study quantum entanglement in condensed matter systems. This cross-pollination enriches both fields; condensed matter techniques, in turn, are utilized to analyze problems in black hole physics, including the black hole information paradox and entropy calculations. Such interdisciplinary endeavors not only widen the scope of string theory applications but also enhance our understanding of quantum mechanics and statistical physics.

One particularly promising direction is the exploration of topological phases of matter through string theoretical frameworks. Topological insulators, which exhibit uniquely protected surface states that are robust against external perturbations due to their topological nature, can similarly be studied using topological concepts from string theory. This alliance might be crucial in identifying new materials with novel properties, potentially leading to revolutionary advances in electronics and quantum computing.

As we advance these interdisciplinary approaches, adopting methodologies like machine learning from the broader field of computer science provides another layer of depth. Techniques such as tensor network simulations have been employed to investigate the entanglement entropy in various quantum states, a concept that is crucial both to quantum information theory and to understanding the fabric of space-time in string theory.

Visual aid suggestion: Integrate a diagram that illustrates the dualities between cosmological phenomena and condensed matter applications—showing, for instance, how tools developed for black hole entropy might be applied to complex problems in superconductivity.

By merging concepts from cosmology, condensed matter physics, and beyond, string theory is not only tested and refined but also provides tools that drive forward understanding in multiple fields. This integrated approach does not merely push the frontiers of individual disciplines but fosters a holistic view of the physical world, allowing us to both pose and answer questions that lie at the very heart of reality itself. Thus, as we continue to weave together insights from varied fields of study, we enrich the overarching framework of string theory while illuminating the dark corners of both microcosmic and macrocosmic realms.

10.9 Growing Global Collaborations and Projects in String Theory

The landscape of string theory, due to its comprehensive demands on both theoretical frameworks and experimental validation, necessitates a broad-based collaborative effort that spans continents and cultures. These collaborations are not merely beneficial; they are imperative for the advancement of the field. As we delve into the present era where complex physical phenomena at both cosmic and quantum scales are being explored, the value of global partnerships has never been more pronounced.

At the heart of these international endeavors is the shared goal to unravel the mysteries of the universe through the lens of string theory. One pivotal aspect of these collaborations is the fusion of diverse theoretical insights which allows for a richer and more diverse intellectual ecosystem. Researchers from various backgrounds bring unique perspectives which contribute to more robust theoretical advancements. For instance, collaborations between theorists in Europe and Asia have recently proposed new models of string compactifications, potentially paving the way for resolving some of the long-standing anomalies in particle physics.

Moreover, with the recent surge in technological advancements, these collaborations have also been greatly facilitated by digital communication platforms that allow for seamless interaction among scientists, regardless of their physical location. Virtual seminars and workshops have become commonplace, enabling an ongoing dialogue among the world's leading string theorists. This democratization of communication ensures that even those situated in remote parts of the world can contribute to

10.9. GROWING GLOBAL COLLABORATIONS AND PROJECTS IN STRING THEORY

and benefit from the global discourse on string theory.

On the experimental front, large-scale projects such as the Large Hadron Collider (LHC) at CERN involve scientists from over a hundred countries and have components sourced from all over the globe. This monumental scientific instrument represents one of many collaborative efforts aiming to provide empirical evidence that could support or refute various aspects of string theory, particularly through the search for supersymmetric particles. Furthermore, upcoming projects like the Square Kilometre Array (SKA) will rely on international cooperation to pioneer new approaches to astronomical observation that could reveal primordial gravitational waves predicted by certain string theory models.

Additionally, interdisciplinary collaboration between string theorists and experts in quantum computing has taken shape. These cooperative efforts aim to utilize quantum algorithms to solve complex string theory problems that are intractable for classical computers. Such synergistic relationships not only accelerate theoretical research but also boost the development of quantum technology, thereby creating a reciprocal framework for progress across different scientific realms.

The educational sphere also reflects this trend towards globalization in string theory research. Numerous universities and institutions worldwide have developed exchange programs and joint research initiatives that allow students and scholars to gain diverse experiences and insights. These educational partnerships are crucial in training the next generation of physicists who will continue to explore the profound questions at the heart of string theory.

Through these multilayered collaborative efforts, string

theory is vigorously being pushed forward on multiple fronts. By weaving together the fabric of worldwide scientific talent and resources, the string theory community is more equipped than ever to tackle the grand challenges of contemporary physics. With every collaborative project and global interaction, we collectively inch closer to understanding the enigmatic universe we inhabit.

10.10 Long-term Outlook: A Vision for String Theory in the Next Century

As we gaze into the distant horizon of string theory, a tapestry of possibilities unfolds, drawing from current research strengths and imaging future landscapes of discovery and application. Central to this vision is the maturation of string theory into a fully integrated theory of quantum gravity, potentially unraveling the deepest mysteries of black holes, the early universe, and perhaps even the very fabric of reality itself.

The convergence of string theory with other theoretical frameworks such as loop quantum gravity or M-theory is likely to yield new mathematical structures and enriched understandings of spacetime and matter. These theoretical advances will require innovations in algebraic geometry, topology, and noncommutative geometry. As these fields deepen and cross-fertilize, the next century of string theory may see a revolution in the ways we comprehend dimensionalities and the interconnectedness of space and time.

10.10. LONG-TERM OUTLOOK: A VISION FOR STRING THEORY IN THE NEXT CENTURY

Experimentally, the expansion of observational strategies will be critical. Advances in telescope technologies and satellite missions could potentially provide indirect evidence of string theory mechanisms, such as the imprints of cosmic strings in the cosmic microwave background. Moreover, experiments in particle physics, sensitive to energies beyond the current reach of the Large Hadron Collider, could test predictions derived from string theory, such as the existence of supersymmetric partners and extra dimensions.

On the technology front, the implications of string theory could permeate to other domains, influencing material science, quantum computing, and even everyday technological devices. The theoretical insights from stringy corrections to quantum mechanics might lead to more robust quantum computers or novel materials with unique electromagnetic properties, conceived from the theories underlying extra dimensions and fundamental strings.

Automation and artificial intelligence (AI) will play increasingly prominent roles in theoretical physics. AI-driven algorithms could handle complex calculations and pattern recognition tasks in data analysis, accelerating the testing of string theoretical models against experimental data. This synergy between advanced computation and theoretical exploration might shorten the feedback loop between hypothesis and verification in string theory research.

Education and public engagement will be imperative in cultivating the next generation of theorists and experimentalists who will carry forward the torch of string theory. Enhanced educational programs that promote the interdisciplinary nature of string theory, blending physics, mathematics, and computation, will be crucial. Public outreach will also be important in demystifying string theory and conveying its implications and potential applications to broader audiences.

The potential for startup companies and new industries based on high-dimensional math and novel materials indicates that string theory may have a practical impact on the economy, fostering new job opportunities and economic growth. Thus, in addition to its profound scientific implications, string theory could be at the heart of a new technological revolution.

Efforts to extend global collaborations will also intensify, as the complex challenges of string theory transcend individual nations or institutions. This shared quest can lead to a more connected and cooperative global scientific community, tackling universally profound questions in a unified effort.

As we look to the future, it is conceivable that string theory will not only deepen our understanding of the universe but also transform various sectors of society through its theoretical insights and applications. It allows us to dream not just of new particles or quantum states, but of new ways in which the world could operate both scientifically and socially.

www.ingramcontent.com/pod-product-compliance
Lightning Source LLC
Chambersburg PA
CBHW052143220526
45471CB00004B/1503